Effects of Plant Extracts on Meat Quality, Intestinal Microbiota and Resistance to Diseases and Stresses of Food Animals

Effects of Plant Extracts on Meat Quality, Intestinal Microbiota and Resistance to Diseases and Stresses of Food Animals

Guest Editors

Shengchen Wang
Yun Hu

MDPI

Basel • Beijing • Wuhan • Barcelona • Belgrade • Novi Sad • Cluj • Manchester

Guest Editors

Shengchen Wang
College of Animal Science and Technology
Yangzhou University
Yangzhou
China

Yun Hu
College of Animal Science and Technology
Yangzhou University
Yangzhou
China

Editorial Office
MDPI AG
Grosspeteranlage 5
4052 Basel, Switzerland

This is a reprint of the Special Issue, published open access by the journal *Animals* (ISSN 2076-2615), freely accessible at: https://www.mdpi.com/journal/animals/special_issues/2GBMT62SG1.

For citation purposes, cite each article independently as indicated on the article page online and as indicated below:

Lastname, A.A.; Lastname, B.B. Article Title. *Journal Name* **Year**, *Volume Number*, Page Range.

ISBN 978-3-7258-5495-0 (Hbk)
ISBN 978-3-7258-5496-7 (PDF)
https://doi.org/10.3390/books978-3-7258-5496-7

Contents

animals

MDPI

Article

The Adaptive Alternation of Intestinal Microbiota and Regulation of Host Genes Jointly Promote Pigs to Digest Appropriate High-Fiber Diets

Yunchao Zhang [1,2], Hui Li [3], Bengao Li [3], Jiayi He [1,2], Chen Peng [1,2], Yanshe Xie [1,2], Guiqing Huang [1,2], Pengju Zhao [1] and Zhengguang Wang [1,2,*]

[1] Hainan Institute, Zhejiang University, Sanya 572000, China; 22117077@zju.edu.cn (Y.Z.); 22017074@zju.edu.cn (J.H.); 22217072@zju.edu.cn (C.P.); 12217021@zju.edu.cn (Y.X.); 22217094@zju.edu.cn (G.H.); zhaopengju2014@zju.edu.cn (P.Z.)

[2] College of Animal Sciences, Zhejiang University, Hangzhou 310000, China

[3] Long Jian Animal Husbandry Company, Haikou 570100, China; 18668180372@163.com (H.L.); 3170100171@zju.edu.cn (B.L.)

* Correspondence: wzhguang68@zju.edu.cn

Simple Summary: Tunchang pigs are known for their good performance on roughage tolerance, which may alleviate the conflict between humans and livestock for grain resources. This study observed that Tunchang pigs can adapt to a high-fiber diet containing 5% crude fiber. The adaptive alterations in intestinal microbiota and the regulation of genes collectively facilitated the digestion of high-fiber diets in the host. These findings provide insight into the roughage tolerance of Tunchang pigs, and contribute to further exploration of host–microbiota interactions.

Abstract: Although studies have revealed the significant impact of dietary fiber on growth performance and nutrient digestibility, the specific characteristics of the intestinal microbiota and gene regulation in pigs capable of digesting high-fiber diets remained unclear. To investigate the traits associated with roughage tolerance in the Chinese indigenous pig breed, we conducted comparative analysis of growth performance, apparent fiber digestibility, intestinal microbiota, SCFA concentrations and intestinal transcriptome in Tunchang pigs, feeding them diets with different wheat bran levels. The results indicated that the growth performance of Tunchang pigs was not significantly impacted, and the apparent total tract digestibility of crude fiber was significantly improved with increasing dietary fiber content. High-fiber diets altered the diversity of intestinal microbiota, and increased the relative abundance of *Prevotella*, *CF231*, as well as the concentrations of isobutyrate, valerate and isovalerate. The LDA analysis identified potential microbial biomarkers that could be associated with roughage tolerance, such as *Prevotella stercorea*, and *Eubacterium biforme*. In addition, appropriate high-fiber diets containing 4.34% crude fiber upregulated the mRNA expressions of PYY, AQP8, and SLC5A8, while downregulating the mRNA expressions of CKM and CNN1.This indicated that appropriate high-fiber diets may inhibit intestine motility and increase the absorption of water and SCFAs.

Keywords: fibrous diets; gut microbiome; roughage tolerance; Tunchang pigs; transcriptome

1. Introduction

The rising demand for feed grain exacerbates the conflict between humans and livestock for grain resources. Therefore, growing attention has been paid to the utilization of agricultural byproducts as feedstuff, such as rice bran and wheat bran, while previous studies found that the addition of fiber-rich ingredients can inhibit nutrient digestibility and reduce feed intake in growing pigs [1–3]. These studies have revealed the significant impact of dietary fiber on growth performance and nutrient digestibility. However, there are certain limitations, as they do not investigate the underlying interactions.

Trillions of microorganisms colonizing the host's large intestine form a complex microbial ecosystem, which is critical for nutrient absorption [4]. The digestive enzymes secreted by the gastrointestinal tract of the pig cannot degrade dietary fiber. Therefore, nondigestible dietary fiber is fermented in the large intestine to support microbial growth. In return, the microbiota generate multiple metabolites, among which short-chain fatty acids (SCFAs) are primary [5]. SCFAs serve as energy substrates, mitigate inflammation, and regulate satiety [6]. A recent study found that the growth performance of weaned piglets fed dietary corn bran or wheat bran improved due to alterations in intestinal microbiota and enhanced butyrate production [7]. Many studies have focused on the impact of dietary fiber on intestinal microbiota and microbial metabolites, overlooking the regulation of dietary fiber on host gene expression, which influences the host's physiological activities.

Chinese indigenous pig breeds have demonstrated good performance in roughage tolerance [8–10], which may be related to intestinal microbiota and host–microbiota interactions [11]. Tunchang pigs, a subpopulation of Hainan pigs, thrive in an environment similar to Southeast Asia, characterized by a mild climate and abundant rainfall. These pigs gradually developed resistance to roughage through ample supplementation of green feed [12]. However, the roughage tolerance of Tunchang pigs remains unknown, and there is a lack of studies investigating the underlying host–microbiota interactions. The present study aimed to investigate the interaction between intestinal microbiota and gene expression in response to feeding high-fiber diets to Tunchang pigs through 16S rRNA gene sequencing and transcriptome sequencing. It was hypothesized that Tunchang pigs would perform well in roughage tolerance and that dietary fiber would affect the composition and function of cecal and colonic microbiota, involving the regulation of intestinal gene expression.

2. Materials and Methods

2.1. Animals, Diets and Sampling

A total of 24 Tunchang female pigs (day 180) with body weights of 43.71 ± 2.89 kg were randomly allotted into three groups: A, B and C group. Each pig was considered a replicate, resulting in 8 replicates per group. Group A was fed a basal diet, while Group B and C were provided with basal diets supplemented with 8% and 16% wheat bran, respectively. The basal diet was formulated to meet the nutrient requirements of the NRC. According to the "unique difference principle", the dietary fiber level in each group was adjusted by manipulating the wheat bran content. Additionally, the contents of corn, soybean meal, grease meal, and fish meal were slightly adjusted to ensure the measured values of crude protein (CP) and digestible energy (DE) were nearly identical (Table 1). All pigs had ad libitum access to their diets and clean water during the 111-day experiment. Feed consumption was recorded daily to calculate the average daily feed intake (ADFI). Each pig was weighed at the beginning and end of the trial to determine the average daily gain (ADG).

Table 1. Ingredients and nutrient level of the experimental diets (as-fed basis).

Items	Group A	Group B	Group C
Ingredient (%)			
Corn	60.95	56.50	52.00
Soybean meal	28.00	25.30	24.10
Wheat bran [1]	2.00	9.50	15.50
Grease meal	3.00	2.80	2.60
Fish meal	2.00	1.85	1.75
Antifungal agent [2]	0.05	0.05	0.05
Premix [3]	4.00	4.00	4.00

Table 1. *Cont.*

Items	Group A	Group B	Group C
Nutrient level [4]			
CP (%)	16.54	16.36	16.20
CF (%)	3.57	4.34	5.00
DE (MJ/KG)	15.10	15.09	15.07

[1] Wheat bran: 14.11% CP, 13.99% CF, 0.15% AIA. [2] Antifungal agent is a feed mold inhibitor that effectively prevents the harm of various mycotoxins to pigs. [3] Premix supplied per kg diet as follows: vitamin A 8000 IU, vitamin D_3 1500 IU, vitamin E 100 mg, vitamin K_3 4 mg, vitamin B_1 2 mg, vitamin B_2 8 mg, vitamin B_6 3 mg, vitamin B_{12} 0.04 mg, niacin 30 mg, pantothenie acid 35 mg, folic acid 0.6 mg, biotin 0.13 mg, Choline 150 mg, Fe 60 mg, Cu 5 mg, Zn 60 mg, Mn 10 mg, Se 0.15 mg and I 0.1 mg. [4] The DE is calculated, whereas all other values are analyzed.

At the end of the trial (day 111), fresh fecal samples were collected to determine nutrient and fiber digestibility. Five pigs from each group were taken to a slaughterhouse, euthanized by electrical stunning and exsanguination after approximately 12 h of fasting. Digesta from the cecum and colon were collected for 16S rRNA gene sequencing and SCFAs analysis. Sterile ophthalmic scissors were used to excise the cecal and colonic segments for transcriptome sequencing.

2.2. Chemical Analysis

The apparent total tract digestibility (ATTD) was determined by using acid-insoluble ash (AIA) as an endogenous indicator. Diets and fecal samples were analyzed for DE, CP, CF and AIA according to the Association of Official Analytical Chemists procedures (AOAC) [13]. ATTD was calculated as follows:

$$\text{ATTD}(\%) = 100\% \times (1 - (\text{NC}_\text{F} \times \text{AIA}_\text{D})/(\text{NC}_\text{D} \times \text{AIA}_\text{F}))$$

where NC_F is the nutrient content in feces; AIA_D is the content of AIA in diets; NC_D is the nutrient content in diets; and AIA_F is the content of AIA in feces.

Cecal and colonic SCFA concentrations were determined by a gas chromatography–tandem mass spectrometry platform at Personal Biotechnology Co., Ltd. (Shanghai, China), according to a previous study [14].

2.3. 16S rRNA Gene Sequencing

DNA extraction and 16S rRNA gene sequencing were conducted at Personal Bio Inc. (Shanghai, China). Microbial DNA was extracted from the digesta of cecum and colon using a DNeasy PowerSoil Kit (QIAGEN, Hilden, Germany,) according to the manufacturer's protocols. The V3-V4 hypervariable regions of the bacteria 16S rRNA gene were amplified with primers 338F and 806R by the T100 PCR system (Bio-Rad, Hercules, CA, USA). Purified amplicons were pooled in equimolar paired-end sequencing performed on an Illumina MiSeq platform (Illumina, San Diego, CA, USA), according to the standard protocols [15]. Raw sequenced reads were processed to obtain high-quality clean reads as described by a previous study [16]. Clean reads were demultiplexed and constructed to an amplicon sequence variant (ASV) feature table using the QIIME 'DADA2' package [17]. Greengenes database (Release 13.8) [18] was used to predict taxonomic assignments for each ASV sequence. Results were then exported for further analysis in R software (v4.1.0).

2.4. Transcriptome Sequencing

Total RNA was extracted from all intestinal segment samples using a Total Kit II (OMEGA, Norcross, GA, USA). The TruSeq RNA Library Prep Kit v2 (Illumina, San Diego, CA, USA) was used to construct cDNA libraries according to the manufacturer's instructions. After RNA extraction, purification, library construction, and PCR, the cDNA libraries were subjected to paired-end sequencing using Illumina NovaSeq 6000 sequencing platform at Personal Bio Inc. (Shanghai, China). The raw data downloaded from platform contains some adapter-contaminated and low-quality reads, which may interfere with subsequent data analysis. Cutadapt (v1.2.1) was used to further filter the raw data to obtain clean data [19]. The clean data was aligned to the reference genome (Sus scrofa 11.1) using the improved BWT algorithm with the HISAT2 software, which offers higher speed and lower resource consumption [20]. The "DESeq2" package in R software (v4.1.0) was used to calculate the fold change in expression between two groups and q-values for multiple testing. Differential expression genes (DEGs) were analyzed and filtered based on the criteria: | log2FoldChange | > 1, and a significant q-value < 0.05.

2.5. Quantitative Real-Time PCR

Total RNA was isolated from cecal tissue by using SPARKeasy Plus Universal Kit (SparkJade Bio Inc., Shandong, China), according to the manufacturer's protocols. Concentration and purity of RNA samples were determined on a NanoDrop 2000 spectrophotometer (Thermo Scientific, Waltham, MA, USA). Single-stranded cDNA was synthesized from 1 μg of total RNA by using the SPARKscript RT Plus Kit (AG0304; SparkJade Bio Inc., Shandong, China), according to the manufacturer's protocols. Quantitative real-time PCR was carried out with the use of 2 × SYBR Green qPCR Mix (AH0104; SparkJade Bio Inc., Shandong, China). The reaction system of quantitative real-time PCR was 10 μL including 5 μL SYBR mix, 1 μL cDNA, 0.2 μL forward and reverse primers (10 μmol/L), and 3.6 μL RNase free water. Quantitative real-time PCR was performed on Bio-Rad CFX96 Real-Time PCR System (Bio-Rad, Hercules, CA, USA). The expression of the genes was calculated relative to the expression of the housekeeping gene β-actin with the formula $2-\Delta\Delta Ct$. Amplification of specific transcripts was confirmed by melting-curve profiles at the end of each PCR. The primer sequences are listed in Supplementary Materials: Table S1.

2.6. Bioinformatics and Statistical Analysis

All data were presented as the mean ± SEM. Comparing the differences of non-normally various indicators among groups was conducted using by the Kruskal–Wallis one-way ANOVA with the Benjamini–Hochberg false discovery rate multiple-testing correction. The differences were considered significant when $p < 0.05$ and a trend when the p value was between 0.05 and 0.10. Correlations between intestinal microbiota and SCFAs were analyzed using Spearman's correlation analysis with the "pheatmap" package in R (v4.1.0).

3. Results

3.1. Growth Performance Was Not Adversely Affected by Dietary Fiber Level

The ADG of pigs in group A, B and C was 0.36 ± 0.10 kg/d, 0.31 ± 0.08 kg/d and 0.32 ± 0.11 kg/d, respectively. The feed conversion rate in each group was 4.72 ± 0.61, 4.94 ± 0.65 and 5.08 ± 0.73, respectively. No significant differences were observed among groups in the growth performance of pigs (Table 2).

Table 2. Responses of growth performance of Tunchang pigs to dietary fiber levels.

Items	Group A	Group B	Group C	*p*-Value
Initial weight (kg)	41.63 ± 2.74	45.75 ± 1.10	43.75 ± 3.14	0.38
Final weight (kg)	82.81 ± 13.17	80.86 ± 9.56	80.44 ± 14.43	0.93
ADG (kg/d)	0.36 ± 0.10	0.31 ± 0.08	0.32 ± 0.11	0.60
ADFI (kg/d)	1.68 ± 0.34	1.51 ± 0.22	1.59 ± 0.35	0.60
F/G	4.72 ± 0.61	4.94 ± 0.65	5.08 ± 0.73	0.62

3.2. Fiber Apparent Digestibility Significantly Improved with Dietary Fiber Increasing

The ATTD of CP and CF were determined (Table 3). As dietary fiber increased, no significant differences in the ATTD of CP (one-way ANOVA, $p = 0.72$) was observed among group A (CP: 0.86 ± 0.01), group B (CP: 0.87 ± 0.03), and group C (CP: 0.87 ± 0.02). However, a high-fiber diet significantly increased the ATTD of CF, with values of 0.59 ± 0.07, 0.73 ± 0.03, and 0.76 ± 0.01 for groups A, B, and C, respectively ($p < 0.05$).

Table 3. Responses of the ATTD of CP and CF to dietary fiber levels.

Items	Group A	Group B	Group C	p-Value
CP	0.86 ± 0.01	0.87 ± 0.03	0.87 ± 0.02	0.98
CF	0.59 ± 0.07^{a}	0.73 ± 0.03^{b}	0.76 ± 0.01^{b}	0.00

Kruskal–Wallis one-way ANOVA was conducted; different letters in the same row indicate a signifi- cant difference ($p < 0.05$), while the lack of variation in letters indicates no significant difference ($p > 0.05$).

3.3. High Fiber Level Altered the Diversity of Cecal and Colonic Microbiota

The cecal and colonic microbial structure and composition were revealed by 16S rRNA gene sequencing. A total of 801,735 and 1,008,607 high-quality sequences were generated from cecal and colonic samples, respectively, with an average of 53,449 and 67,240 sequences per sample. The coverage index in this experiment consistently exceeded 0.99, indicating that the sequencing results adequately represent the breadth and depth of the samples. There were no significant differences in species richness (as indicated by the ACE and Chao1 indices) or diversity (as indicated by Shannon and Simpson indices) in the cecal and colonic microbiota at the taxonomic level (Figure 1a,b). However, ANOISM analysis based on Bray–Curtis distance showed that differences in the cecal and colonic microbiota among groups were significantly greater than differences within each group ($p < 0.05$) (Figure 1c,d). This suggested notable variations in the composition of cecal and colonic microbiota in Tunchang pigs under different dietary fiber levels ($p < 0.05$).

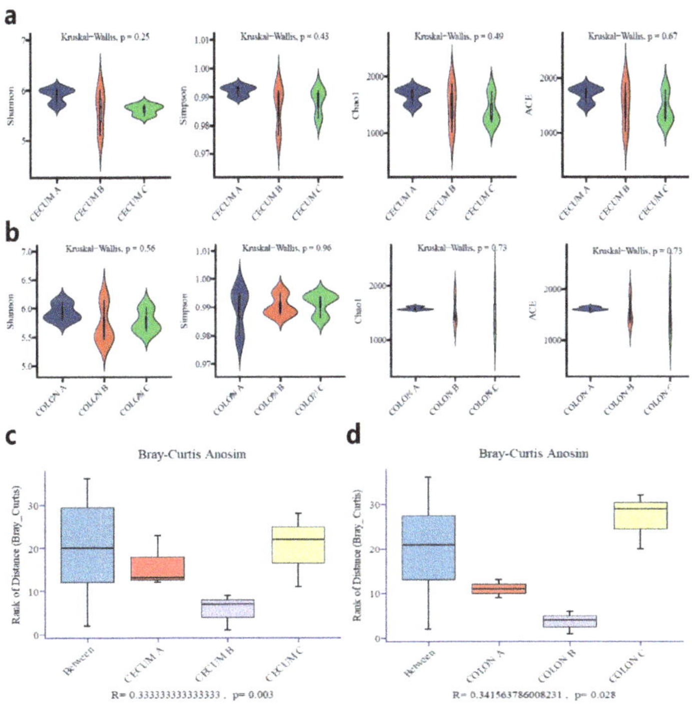

Figure 1. The comparison of microbial diversity of Tunchang pigs among groups. Alpha diversity of cecal microbiota (**a**) and colonic microbiota (**b**). Anoism test of cecal microbiota (**c**) and colonic microbiota (**d**) based on Bray–Curtis distance.

3.4. High Fiber Level Increased the Relative Abundance of Intestinal Microbiota Related to Fiber Degradation

In the cecum, Firmicutes, Bacteroidetes and Proteobacteria were the three dominant phyla (Figure 2a), which accounted for 51.30%, 36.61%, and 6.77% in group A, 42.74%,

46.62%, and 8.13% in group B, and 43.72%, 48.22% and 5.65% in group C, respectively. The relative abundance of Bacteroidetes significantly increased ($p < 0.05$). It is worth noting that there was a significant decreasing trend ($p < 0.10$) in the relative abundance of Firmicutes and Verrucomicrobia with the increase in dietary fiber levels. At the genus level, the 10 most predominant genera (those with a relative abundance ≥3% in at least one group) were Prevotella, unidentified_Bacteroidales, unidentified_Ruminococcaceae, unidentified_S24-7, unidentified_Lachnospiraceae, CF231, Phascolarcto-bacterium, Succinivibrio, and SMB53 (Figure 2b). The relative abundance of CF231 was significantly increased in group C compared with other groups ($p < 0.05$).

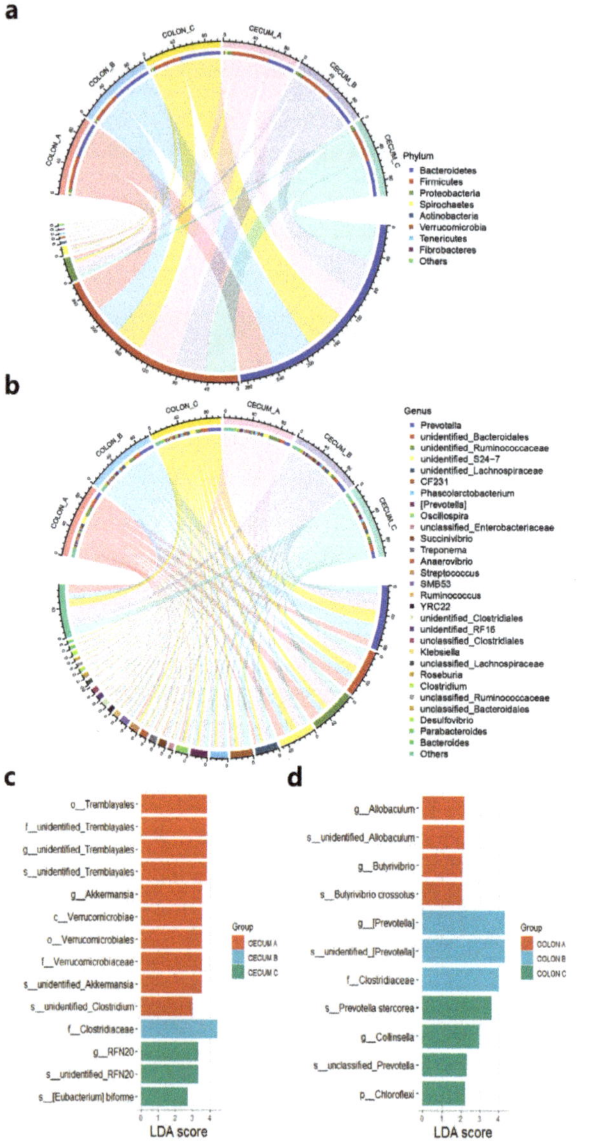

Figure 2. The intestinal microbial composition in the cecum and colon of Tunchang pigs evaluated at the phylum (**a**) and genus (**b**) level. The LDA analysis in the cecum (**c**) and colon (**d**).

In the colon, similar to cecum, the three dominant phyla were Bacteroidetes, Firmicutes, and Proteobacteria (Figure 2a), contributing 46.54%, 44.38% and 4.47% to group A, 55.47%, 39.35% and 1.70% to group B, and 55.06%, 38.30% and 3.81% to group C, respectively. There was a significant increasing trend ($p < 0.10$) in the relative abundance of Bacteroidetes with the increase in dietary fiber levels. At the genus level, the relative abundance of Prevotella significantly increased in group B compared with group A ($p < 0.05$) (Figure 2b).

Some key microbial biomarkers were identified by linear discriminant analysis (LDA) coupled with LDA Effect Size. In the cecum, four microbial biomarkers (LDA scores > 2) were obtained from group B and C, including f_Clostridiaceae, g_RFN20, s_unidentified RFN20, and s_[Eubacterium] biforme (Figure 2c). In the colon, there were seven microbial biomarkers in group B and C, mainly from the Prevotella genus, such as *Prevotella stercorea*, as well as f_Clostridiaceae and g_Collinsella (Figure 2d). These microbial colonies were likely to be associated with a fiber degradation function.

3.5. Dietary Fiber Promotes the Increase in the Functionality Activity of the Microbiota and SCFAs

The alpha diversity analysis and LDA focused on the diversity and composition of intestinal microbiota. Microbial functions of different microbiota among groups were predicted with PICRUSt2 analysis, referring to the MetaCyc and KEGG database. The PICRUSt2 analysis showed that most microbial metabolic pathways were enriched in biosynthesis, degradation, generation of precursor metabolite and energy processes. In terms of biosynthesis, the main focus was on amino acids, cofactors, coenzymes, electron carriers, vitamins, nucleosides and nucleotides, as well as fatty acid and lipid biosynthesis. In terms of degradation, the main focus was on carbohydrate, carboxylate, nucleoside and nucleotide degradation. The most significant processes in the generation of precursor metabolite and energy was fermentation, followed by glycolysis and the tricarboxylic acid cycle (Figure 3a). Similar results were found in the colon. One-way ANOVA analysis found several microbial pathways of KEGG and MetaCyc had higher abundance in group B and C compared to those in group A ($p < 0.05$), such as fructose and mannose metabolism, lipopolysaccharide biosynthesis, and lipid IVA biosynthesis (Figure 3b,c). These enriched microbial pathways helped us understand the active metabolism in cecal and colonic microbiota, and also demonstrated that these microbiota ferment undigested carbohydrates that are not absorbed in the small intestine to produce SCFAs and provide energy.

The SCFA concentrations in the cecum and colon were next determined using GC-MS. One-way ANOVA revealed that the cecal concentrations of isobutyrate, valerate and isovalerate significantly increased ($p < 0.05$) in group C compared with other groups (Figure 3d). Dietary fiber level did not affect the acetate, propionate and butyrate concentrations in the cecum. The colonic SCFA concentrations were not affected by dietary fiber level.

3.6. Differentially Expressed Genes Related to Roughage Tolerance in the Cecum

Transcriptome sequencing was employed to examine the impact of dietary fiber levels on mRNA expression of genes in the cecum. Enrichment analysis of DEGs revealed that most of these genes are enriched in pathways related to cell signaling and cell adhesion (Figure 4a). Notably, the mRNA expressions of PYY and AQP8 were significantly increased in the cecum of pigs in group B compared with group A, whereas the mRNA expressions of CKM and CNN1 were significantly decreased ($p < 0.05$) (Figure 4b). Additionally, The mRNA expression of SLC5A8 was significantly elevated in the cecum of pigs fed high-fiber diets ($p < 0.05$). The results of quantitative real-time PCR corroborated the findings from the transcriptome analysis (Figure 4c).

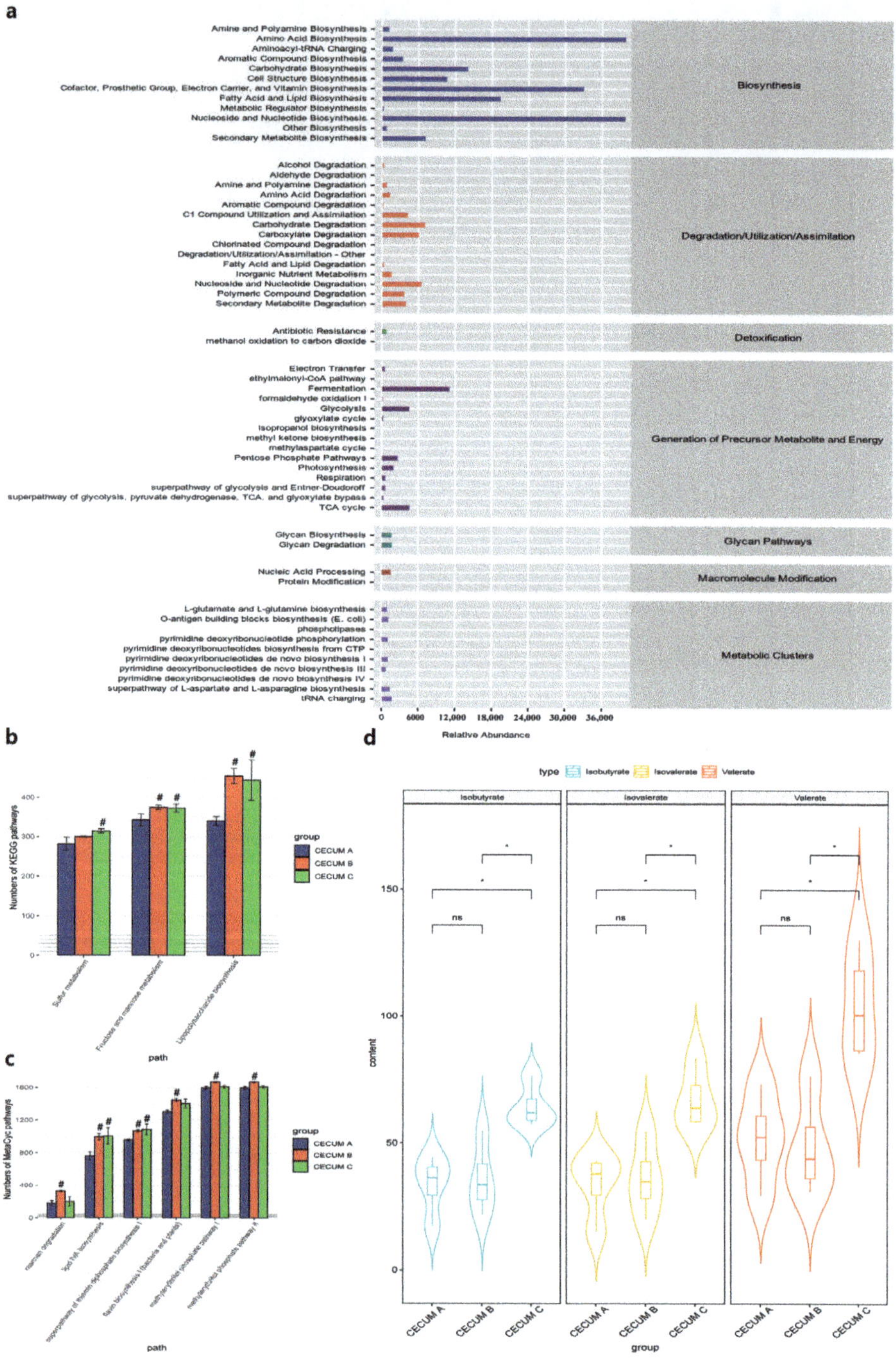

Figure 3. The predicted functions of cecal microbiota of Tunchang pigs based on PICRUSt analysis (**a**). Significantly changed microbial functions of KEGG (**b**) and MetaCyc (**c**). $^{\#}$ $p < 0.05$ compared with group A. Significantly changed SCFAs (**d**). * $p < 0.05$; ns, not significant.

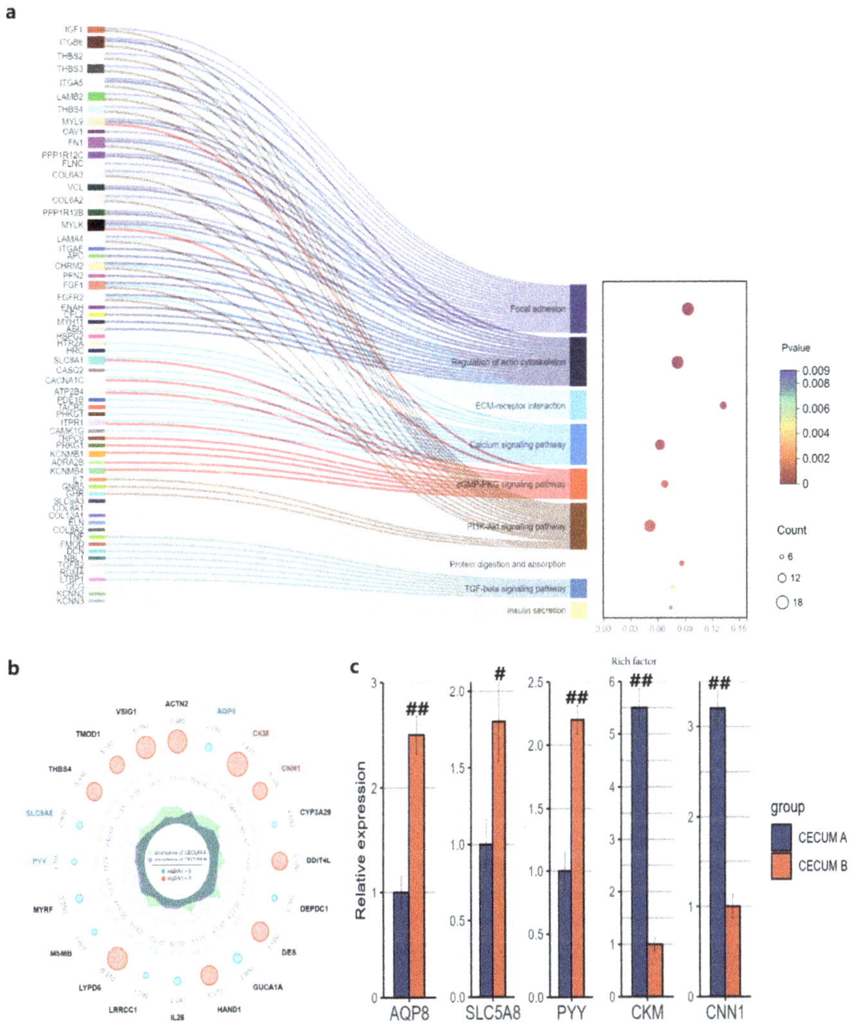

Figure 4. The top 20 differentially expressed genes between group A and B (**a**) in the cecum of Tunchang pigs. KEGG enrichment analysis of differentially expressed genes (**b**). Relative expression of several genes as determined by quantitative real-time PCR (**c**). AQP8, aquaporin 8; SLC5A8, solute carrier family 5 member 8; PYY, peptide YY; CKM, creatine kinase, M-type; CNN1, calponin 1. # $p < 0.05$, ## $p < 0.01$.

4. Discussion

Dietary fiber is essential for the normal physiological metabolism of monogastric animals, as revealed by studies on its physicochemical properties and nutritional functions [21]. Our results showed no significant difference in the growth performance of pigs fed high-fiber diets, indicating that Tunchang fattening pigs could tolerate the diet with 5% CF. Similar results have been reported previously [9,22,23]. However, there were some contradictory results in the previous studies. For example, the inclusion rate of distiller dried grains with solubles was increased from 0 to 20% in nursery pig diets [24] and from 0 to 30% in growing-pig diets [25]. This resulted in a linear decrease in pig body weight, despite the diets being formulated to contain similar net energy. Various factors, such as genetic background and husbandry management can lead to different outcomes in similar studies, both of which affect growth performance. Tunchang pigs have gradually developed good performance on roughage tolerance through long-term breeding [12].

The roughage tolerance of Tunchang pigs was demonstrated by the significant improvement in the ATTD of CF with increased dietary fiber. The ATTD determined using AIA as an indigestible marker was highly correlated with that determined using total fecal collection [26], providing a more intuitive understanding of the digestion and absorption of nutrients in pigs. Compared with other animal trials [27,28], the ATTD of CP and CF determined in the present study were slightly higher. We speculated that pigs ingested a small amount of dust ash during feeding and drinking, leading to higher AIA content in feces. Given the small within-group error, the ATTD reflected the nutrient digestion and absorption in pigs.

Due to pigs' inability to secrete fiber-digesting enzymes, dietary fiber degradation primarily relies on fermentation by the cecal and colonic microbiota. Studies have shown that diet plays a crucial role in regulating the composition of intestinal microbiota [29], with dietary fiber being a key driver of these changes [30]. Changes in dietary fiber can alter the intestinal microbiota's metabolic pathways and reorganize the microbial community in order to adapt to the new diet. In this study, high-fiber diets significantly increased the relative abundance of Bacteroidetes in the cecum, with a significant increasing trend in the colon. Bacteroidetes are dominant symbiotic bacteria in the intestines, with a sophisticated capacity to degrade polysaccharides and extract energy [31]. Additionally, Xie et al. [32] speculated that the higher abundance of Bacteroidetes in the intestines of Laiwu pigs may be related to the excellent roughage tolerance of Chinese indigenous pig breeds. A significant decreasing trend was observed in the relative abundance of Verrucomicrobia in cecal and colonic microbiota with increased dietary fiber. Similarly, a previous study reported a significant decrease in the relative abundance of Verrucomicrobia when Heigai pigs were fed a diet containing 30% alfalfa [33]. At the genus level, *Prevotella* was consistently one of the dominant genera, regardless of the dietary fiber level and intestinal segment. This finding is in line with previous studies [34–36], indicating that *Prevotella* may be a signature genus in intestinal microbiota of pigs. Meanwhile, the relative abundance of *Prevotella* in the colon significantly increased with the increase in dietary fiber. As a genus of Bacteroidetes, *Prevotella* exhibits diverse capabilities in utilizing complex carbohydrates [37]. In the cecum, *CF231* had a high relative abundance, and its relative abundance significantly increased with the increase in dietary fiber, especially in group C, accounting for 11.05%. The *CF231* is commonly present in the rumen of ruminants, and is one of the dominant genera [38,39]. Huang et al. [40] found a significant positive correlation between *CF231* and carbohydrate metabolism in the functional prediction of rumen microbiota in dairy cows. In a study on the differences of colonic microbiota and microbial metabolism between Landrace and Meishan pigs, correlation analysis showed a significant positive correlation between *CF231* and SCFAs [41].

To understand the differences in intestinal microbiota associated with high or low ATTD of CF, LDA was used to identify potential microbial biomarkers related to roughage tolerance, such as *Prevotella stercorea* and *Eubacterium biforme*. The PICRUSt2 accurately predicts the functional potential of microbial communities based on 16S rRNA sequencing data [42]. In this study, PICRUSt2 analysis revealed that the cecal and colonic microbiota primarily ferment CF to produce SCFAs for energy. This underscores the crucial role of intestinal microbiota in the host's roughage tolerance. There were higher cecal concentrations of isobutyrate, valerate and isovalerate in group C compared with the other two groups, while there was no significant difference in the colonic SCFA concentrations among groups. Subsequent correlation analysis revealed the main relationships between intestinal microbiota and SCFAs, explaining the significant differences in the cecal SCFA concentrations.

SCFAs are an important source of energy for the host and also regulate host gene expression, thereby modulating physiological activities [43]. Transcriptome sequencing was used to identify DEGs in the cecum of Tunchang pigs among three groups with different dietary fiber level. It could help us further explore candidate genes and pathways that related to roughage tolerance of Tunchang pigs. In this study, enrichment analysis of DEGs revealed that most of these genes are involved in pathways related to cell signaling and cell adhesion. Notably, the mRNA expression of PYY was significantly increased in the cecum of pigs in group B. PYY encodes peptide YY, a short peptide released by intestinal cells that regulate satiety. By binding to neuropeptide Y receptors, peptide YY inhibits gastrointestinal motility and increases water absorption [44]. Correspondingly, there was significant downregulation of CKM and CNN1, which are involved in regulating smooth muscle contraction, and significant upregulation of AQP8, encoding aquaporin proteins. Furthermore, a recent study indicated that peptide YY also functions as an antimicrobial peptide, capable of inhibiting the transformation of virulent Candida albicans and selectively promoting the proliferation of non-pathogenic yeast forms. It plays a crucial role in maintaining fungal balance in the mammalian digestive system [45]. Additionally, the mRNA expression of SLC5A8 was significantly increased in the cecum and colon of pigs fed with high-fiber diets. SLC5A8, also known as SMCT1, encodes sodium-coupled monocarboxylate transporter 1, which mediates the active transport of SCFAs within the intestine, facilitating their absorption by intestinal epithelial cells [46]. Meanwhile, a study in mice revealed that SLC5A8 inhibits the differentiation of naive T cells into pro-inflammatory IFN γ-secreting cells by transporting butyrate. It has also been observed that SLC5A8 can prevent colitis and colon cancer under conditions of low fiber intake. It is concluded that SLC5A8 serves as a crucial link between dietary fiber and the mucosal immune system through the bacterial metabolite butyrate [47]. Similarly, the upregulation of PYY and SLC5A8 was also shown in the cecum of pigs fed 15% dietary alfalfa meal [48]. We speculated that appropriate high-fiber diets could inhibit intestine motility and increase the absorption of water and SCFAs.

5. Conclusions

In conclusion, this study observed that Tunchang pigs can adapt to a high-fiber diet with 5% crude fiber. The adaptative alternation of intestinal microbiota and regulation of genes jointly promoted the host to digest appropriate high-fiber diets. These findings draw insight into the roughage tolerance of Tunchang pigs, and help in further exploring host–microbiota interactions.

Supplementary Materials: The following supporting information can be downloaded at: https://www.mdpi.com/article/10.3390/ani14142076/s1, Table S1: The primer sequences used in quantitative real-time PCR.

Author Contributions: Y.Z.: Conceptualization, methodology, data curation, visualization, writing original draft. H.L.: Conceptualization, methodology, funding acquisition. B.L.: Methodology, funding acquisition. J.H.: Data curation, writing review. C.P., Y.X., G.H. and P.Z.: Supervision, conceptualization. Z.W.: Supervision, conceptualization, funding acquisition. All authors have read and agreed to the published version of the manuscript.

Funding: The research was founded by the Hainan Province Science and Technology Special Fund (ZDYF2022XDNY238) and the Project of Sanya Yazhou Bay Science and Technology City (SCKJ-JYRC-2022-07).

Institutional Review Board Statement: The study was conducted according to the guidelines of the Declaration of Helsinki, and approved by the Laboratory Animal Welfare and Ethics Committee of Zhejiang University (protocol code: ZJU20230008, 20 March 2023).

Informed Consent Statement: Not applicable.

Data Availability Statement: The datasets used or analyzed during the current study are available from the corresponding author on reasonable request.

Conflicts of Interest: Hui Li and Bengao Li are employed at Long Jian Husbandry Company. This organization played no role in sponsoring or product advertising. Other authors declare no financial interest or relevant conflicts of interest.

References

1. Cervantes-Pahm, S.K.; Liu, Y.; Evans, A.; Stein, H.H. Effect of novel fiber ingredients on ileal and total tract digestibility of energy and nutrients in semi-purified diets fed to growing pigs. *J. Sci. Food Agric.* **2014**, *94*, 1284–1290. [CrossRef]
2. Zhang, W.; Li, D.; Liu, L.; Zang, J.; Duan, Q.; Yang, W.; Zhang, L. The effects of dietary fiber level on nutrient digestibility in growing pigs. *J. Anim. Sci. Biotechnol.* **2013**, *4*, 17. [CrossRef]
3. Wilfart, A.; Montagne, L.; Simmins, H.; Noblet, J.; van Milgen, J. Effect of fibre content in the diet on the mean retention time in different segments of the digestive tract in growing pigs. *Livest. Sci.* **2007**, *109*, 27–29. [CrossRef]
4. Yatsunenko, T.; Rey, F.E.; Manary, M.J.; Trehan, I.; Dominguez-Bello, M.G.; Contreras, M.; Magris, M.; Hidalgo, G.; Baldassano, R.N.; Anokhin, A.P.; et al. Human gut microbiome viewed across age and geography. *Nature* **2012**, *486*, 222–227. [CrossRef]
5. Roy, C.C.; Kien, C.L.; Bouthillier, L.; Levy, E. Short-chain fatty acids: Ready for prime time? *Nutr. Clin. Pract.* **2006**, *21*, 351–366. [CrossRef] [PubMed]
6. Koh, A.; Vadder, F.D.; Kovatcheva-Datchary, P.; Bäckhed, F. From Dietary Fiber to Host Physiology: Short-Chain Fatty Acids as Key Bacterial Metabolites. *Cell* **2016**, *165*, 1332–1345. [CrossRef]
7. Zhao, J.; Liu, P.; Wu, Y.; Guo, P.; Liu, L.; Ma, N.; Levesque, C.; Chen, Y.; Zhao, J.; Zhang, J.; et al. Dietary Fiber Increases Butyrate-Producing Bacteria and Improves the Growth Performance of Weaned Piglets. *J. Agric. Food Chem.* **2018**, *66*, 7995–8004. [CrossRef]
8. Gao, Q.; Sun, G.; Duan, J.; Luo, C.; Yangji, C.; Zhong, R.; Chen, L.; Zhu, Y.; Wangdui, B.; Zhang, H. Alterations in gut microbiota improve SCFA production and fiber utilization in Tibetan pigs fed alfalfa diet. *Front. Microbiol.* **2022**, *13*, 969524. [CrossRef] [PubMed]
9. Pu, G.; Li, P.; Du, T.; Niu, Q.; Fan, L.; Wang, H.; Liu, H.; Li, K.; Niu, P.; Wu, C.; et al. Adding Appropriate Fiber in Diet Increases Diversity and Metabolic Capacity of Distal Gut Microbiota without Altering Fiber Digestibility and Growth Rate of Finishing Pig. *Front. Microbiol.* **2020**, *11*, 533. [CrossRef]
10. Urriola, P.E.; Stein, H.H. Comparative digestibility of energy and nutrients in fibrous feed ingredients fed to Meishan and Yorkshire pigs. *J. Anim. Sci.* **2012**, *90*, 802–812. [CrossRef]
11. Pu, G.; Hou, L.; Du, T.; Zhou, W.; Liu, C.; Niu, P.; Wu, C.; Bao, W.; Huang, R.; Li, P. Increased Proportion of Fiber-Degrading Microbes and Enhanced Cecum Development Jointly Promote Host to Digest Appropriate High-Fiber Diets. *mSystems* **2023**, *8*, e00937-22. [CrossRef] [PubMed]
12. China National Commission of Animal Genetic Resources. *Animal Genetic Resources in China: Pigs*; China Agriculture Press: Beijing, China, 2011.
13. AOAC. *Official Methods of Analysis*, 18th ed.; Association of Official Analytical Chemists: Gaithersburg, MD, USA, 2007.
14. He, J.; Zhang, Y.; Li, H.; Xie, Y.; Huang, G.; Peng, C.; Zhao, P.; Wang, Z. Hybridization alters the gut microbial and metabolic profile concurrent with modifying intestinal functions in Tunchang pigs. *Front. Microbiol.* **2023**, *14*, 1159653. [CrossRef] [PubMed]
15. Caporaso, J.G.; Lauber, C.L.; Walters, W.A.; Berg-Lyons, D.; Huntley, J.; Fierer, N.; Owens, S.M.; Betley, J.; Fraser, L.; Bauer, M.; et al. Ultra-high-throughput microbial community analysis on the Illumina HiSeq and MiSeq platforms. *ISME J.* **2012**, *6*, 1621–1624. [CrossRef] [PubMed]
16. Bergamaschi, M.; Tiezzi, F.; Howard, J.; Huang, Y.J.; Gray, K.A.; Schillebeeckx, C.; McNulty, N.P.; Maltecca, C. Gut microbiome composition differences among breeds impact feed efficiency in swine. *Microbiome* **2020**, *8*, 110. [CrossRef] [PubMed]
17. Callahan, B.J.; McMurdie, P.J.; Rosen, M.J.; Han, A.W.; Johnson, A.J.A.; Holmes, S.P. DADA2, High-resolution sample inference from Illumina amplicon data. *Nat. Methods* **2016**, *13*, 581–583. [CrossRef]
18. DeSantis, T.Z.; Hugenholtz, P.; Larsen, N.; Rojas, M.; Brodie, E.L.; Keller, K.; Huber, T.; Dalevi, D.; Hu, P.; Andersen, G.L. Greengenes a chimera-checked 16S rRNA gene database workbench compatible with, A.R.B. *Appl. Environ. Microbiol.* **2006**, *72*, 5069–5072. [CrossRef]
19. Martin, M. Cutadapt removes adapter sequences from high-throughput sequencing reads. *EMBnet J.* **2011**, *17*, 10. [CrossRef]
20. Kim, D.; Paggi, J.M.; Park, C.; Bennett, C.; Salzberg, S.L. Graph-based genome alignment and genotyping with HISAT2 and HISAT-genotype. *Nat. Biotechnol.* **2019**, *37*, 907–915. [CrossRef]
21. Riva, A.; Kuzyk, O.; Forsberg, E.; Siuzdak, G.; Pfann, C.; Herbold, C.; Daims, H.; Loy, A.; Warth, B.; Berry, D. A fiber-deprived diet disturbs the fine-scale spatial architecture of the murine colon microbiome. *Nat. Commun.* **2019**, *10*, 4366. [CrossRef]
22. Wu, F.; Johnston, L.J.; Urriola, P.E.; Hilbrands, A.M.; Shurson, G.C. Evaluation of NE predictions and the impact of feeding maize distillers dried grains with solubles (DDGS) with variable NE content on growth performance and carcass characteristics of growing-finishing pigs. *Anim. Feed Sci. Technol.* **2016**, *215*, 105–116. [CrossRef]
23. Kerr, B.J.; Gabler, N.K.; Shurson, G.C. Formulating diets containing corn distillers dried grains with solubles on a net energy basis: Effects on pig performance and on energy and nutrient digestibility. *Prof. Anim. Sci.* **2015**, *31*, 497–503. [CrossRef]
24. Avelar, E.; Jha, R.; Beltranena, E.; Cervantes, M.; Morales, A.; Zijlstra, R.T. The effect of feeding wheat distillers dried grain with solubles on growth performance and nutrient digestibility in weaned pigs. *Anim. Feed Sci. Technol.* **2010**, *160*, 73–77. [CrossRef]
25. Agyekum, A.K.; Woyengo, T.A.; Slominski, B.A.; Yin, Y.L.; Nyachoti, C.M. Effects of formulating growing pig diet with increasing levels of wheat-corn distillers dried grains with solubles on digestible nutrient basis on growth performance and nutrient digestibility. *J. Anim. Physiol. Anim. Nutr.* **2014**, *98*, 651–658. [CrossRef] [PubMed]
26. Prawirodigdo, S.; Gannon, N.J.; Leury, B.J.; Dunshea, F.R. Acid-insoluble ash is a better indigestible marker than chromic oxide to measure apparent total tract digestibility in pigs. *Anim. Nutr.* **2021**, *7*, 64–71. [CrossRef] [PubMed]

27. Oh, H.J.; Yun, W.; Lee, J.H.; An, J.S.; Kim, T.H.; Cho, E.A.; Park, B.J.; Kim, K.H.; Lee, S.D.; Cho, J.H. Effect of replacing corn with soy hulls on nutrient digestibility of growing pigs. *J. Anim. Sci. Technol.* **2020**, *62*, 180–186. [CrossRef] [PubMed]
28. Liu, J.B.; Liu, Z.Q.; Chen, L.; Zhang, H.F. Effects of feed intake and dietary nutrient density on apparent ileal and total tract digestibility of nutrients and gross energy for growing pigs. *J. Anim. Sci.* **2016**, *94*, 4251–4258. [CrossRef] [PubMed]
29. Burbach, K.; Strang, E.J.P.; Mosenthin, R.; Camarinha-Silva, A.; Seifert, J. Porcine intestinal microbiota is shaped by diet composition based on rye or triticale. *J. Appl. Microbiol.* **2017**, *123*, 1571–1583. [CrossRef]
30. Heinritz, S.N.; Weiss, E.; Eklund, M.; Aumiller, T.; Louis, S.; Rings, A.; Messner, S.; Camarinha-Silva, A.; Seifert, J.; Bischoff, S.C.; et al. Intestinal Microbiota and Microbial Metabolites Are Changed in a Pig Model Fed a High-Fat/Low-Fiber or a Low-Fat/High-Fiber Diet. *PLoS ONE* **2016**, *11*, e0154329. [CrossRef]
31. Sonnenburg, J.L.; Xu, J.; Leip, D.D.; Chen, C.-H.; Westover, B.P.; Weatherford, J.; Buhler, J.D.; Gordon, J.I. Glycan foraging in vivo by an intestine-adapted bacterial symbiont. *Science* **2005**, *307*, 1955–1959. [CrossRef]
32. Xie, C.; Zhu, X.; Xu, B.; Niu, Y.; Zhang, X.; Ma, L.; Yan, X. Integrated analysis of multi-tissues lipidome and gut microbiome reveals microbiota-induced shifts on lipid metabolism in pigs. *Anim. Nutr.* **2022**, *10*, 280–293. [CrossRef]
33. Li, J.; Zhang, S.; Gu, X.; Xie, J.; Zhu, X.; Wang, Y.; Shan, T. Effects of alfalfa levels on carcass traits, meat quality, fatty acid composition, amino acid profile, and gut microflora composition of Heigai pigs. *Front. Nutr.* **2022**, *9*, 975455. [CrossRef] [PubMed]
34. Wang, C.; Wei, S.; Chen, N.; Xiang, Y.; Wang, Y.; Jin, M. Characteristics of gut microbiota in pigs with different breeds, growth periods and genders. *Microb. Biotechnol.* **2022**, *15*, 793–804. [CrossRef] [PubMed]
35. Yang, L.; Wang, G.; Zhou, J.; Yang, Y.; Pan, H.; Zeng, X.; Qiao, S. Exploration of the Potential for Efficient Fiber Degradation by Intestinal Microorganisms in Diqing Tibetan Pigs. *Fermentation* **2021**, *7*, 275. [CrossRef]
36. Cheng, P.; Wang, Y.; Liang, J.; Wu, Y.; Wright, A.; Liao, X. Exploratory Analysis of the Microbiological Potential for Efficient Utilization of Fiber Between Lantang and Duroc Pigs. *Front. Microbiol.* **2018**, *9*, 1342. [CrossRef]
37. Gálvez, E.J.; Iljazovic, A.; Amend, L.; Lesker, T.R.; Renault, T.; Thiemann, S.; Hao, L.; Roy, U.; Gronow, A.; Charpentier, E.; et al. Distinct Polysaccharide Utilization Determines Interspecies Competition between Intestinal *Prevotella* spp. *Cell Host Microbe* **2020**, *28*, 838–852.e6. [CrossRef] [PubMed]
38. Goiri, I.; de Otálora, X.D.; Ruiz, R.; Rey, J.; Atxaerandio, R.; Lavín, J.L.; Martin, D.S.; Orive, M.; Iñarra, B.; Zufia, J.; et al. Spent Coffee Grounds Alter Bacterial Communities in Latxa Dairy Ewes. *Microorganisms* **2020**, *8*, 1961. [CrossRef] [PubMed]
39. Zhang, T.; Mu, Y.; Zhang, D.; Lin, X.; Wang, Z.; Hou, Q.; Wang, Y.; Hu, Z. Determination of microbiological characteristics in the digestive tract of different ruminant species. *Microbiologyopen* **2019**, *8*, e00769. [CrossRef] [PubMed]
40. Huang, S.; Ji, S.; Yan, H.; Hao, Y.; Zhang, J.; Wang, Y.; Cao, Z.; Li, S. The day-to-day stability of the ruminal and fecal microbiota in lactating dairy cows. *Microbiologyopen* **2020**, *9*, e990. [CrossRef]
41. Yan, S.; Zhu, C.; Yu, T.; Huang, W.; Huang, J.; Kong, Q.; Shi, J.; Chen, Z.; Liu, Q.; Wang, S.; et al. Studying the Differences of Bacterial Metabolome and Microbiome in the Colon between Landrace and Meihua Piglets. *Front. Microbiol.* **2017**, *8*, 1812. [CrossRef]
42. Douglas, G.M.; Maffei, V.J.; Zaneveld, J.R.; Yurgel, S.N.; Brown, J.R.; Taylor, C.M.; Huttenhower, C.; Langille, M.G.I. PICRUSt2 for prediction of metagenome functions. *Nat. Biotechnol.* **2020**, *38*, 685–688. [CrossRef]
43. Krautkramer, K.A.; Fan, J.; Bäckhed, F. Gut microbial metabolites as multi-kingdom intermediates. *Nat. Rev. Microbiol.* **2021**, *19*, 77–94. [CrossRef] [PubMed]
44. Cooper, J.A. Factors affecting circulating levels of peptide YY in humans: A comprehensive review. *Nutr. Res. Rev.* **2014**, *27*, 186–197. [CrossRef] [PubMed]
45. Pierre, J.F.; Peters, B.M.; La Torre, D.; Sidebottom, A.M.; Tao, Y.; Zhu, X.; Cham, C.M.; Wang, L.; Kambal, A.; Harris, K.G.; et al. Peptide YY: A Paneth cell antimicrobial peptide that maintains Candida gut commensalism. *Science* **2023**, *381*, 502–508. [CrossRef] [PubMed]
46. Miyauchi, S.; Gopal, E.; Fei, Y.-J.; Ganapathy, V. Functional Identification of SLC5A8, a Tumor Suppressor Down-regulated in Colon Cancer, as a Na^+-coupled Transporter for Short-chain Fatty Acids. *J. Biol. Chem.* **2004**, *279*, 13293–13296. [CrossRef]
47. Gurav, A.; Sivaprakasam, S.; Bhutia, Y.D.; Boettger, T.; Singh, N.; Ganapathy, V. Slc5a8, a Na^+-coupled high-affinity transporter for short-chain fatty acids, is a conditional tumor suppressor in colon that protects against colitis and colon cancer under low-fiber dietary conditions. *Biochem. J.* **2015**, *469*, 267–278. [CrossRef]
48. Wang, J.; Qin, C.; He, T.; Qiu, K.; Sun, W.; Zhang, X.; Jiao, N.; Zhu, W.; Yin, J. Alfalfa-containing diets alter luminal microbiota structure and short chain fatty acid sensing in the caecal mucosa of pigs. *J. Anim. Sci. Biotechnol.* **2018**, *9*, 11. [CrossRef]

animals

MDPI

Review

Regulation Progression on Ellagic Acid Improving Poultry Production Performance by Regulating Redox Homeostasis, Inflammatory Response, and Cell Apoptosis

Shengchen Wang [1,†], Wenjun Zhang [1,†], Bing Tian [1], Yun Hu [1], Tingting Li [1], Xiaoyan Cui [1], Liyang Zhang [2] and Xugang Luo [1,*]

1 College of Animal Science and Technology, Yangzhou University, Yangzhou 225000, China; shengchenwang@yzu.edu.cn (S.W.); 19822629819@163.com (W.Z.)
2 State Key Laboratory of Animal Nutrition, Mineral Nutrition Research Division, Institute of Animal Science, Chinese Academy of Agricultural Sciences, Beijing 100193, China
* Correspondence: wlysz@263.net
† These authors contributed equally to this work.

Simple Summary: Ellagic acid, extracted from medicinal foods like raspberries and pomegranates, enhances poultry production and disease resistance by regulating redox, inflammation, and cell death. However, the research on ellagic acid in poultry is nascent, with unclear mechanisms. This mini-review summarizes the latest 10-year findings to promote the practical use of ellagic acid in maintaining poultry health and formulating strategies.

Abstract: It has been approximately 2000 years since the medicinal homologous theory, which primarily holds that food has the same therapeutic value as medicine in order to improve the health of both humans and animals. In recent years, this theory has also been proposed to be used in poultry breeding. Ellagic acid (EA), a natural compound primarily extracted from medicinal homologous foods such as raspberries and pomegranates, is reported to have incomparable advantages in improving the production performance and disease resistance of poultry due to its pharmacological properties, which regulate the processes of redox homeostasis, inflammatory response, and cell apoptotic death. However, the application and research of EA in poultry production are still in the initial stage, and the potential mechanisms of its biological functions affecting animal health have not been clearly identified, which requires more attention worldwide. This mini-review collects the latest 10-year achievements of research on the effects of EA on poultry health, aiming to promote the practical application of EA in maintaining animal health and formulating corresponding targeted strategies.

Keywords: medicinal homologous; poultry breeding; oxidative stress; inflammation; apoptosis

1. Introduction

For the growth and health of livestock and poultry, oxidative stress, inflammation, and apoptosis occurring within tissue cells are important biological processes that have a significant impact on animal production performance and disease resistance, exhibiting complex interactions that depend on specific production and health environments [1,2]. Oxidative stress, typically characterized by the excessive production of free radicals, such as reactive oxygen species (ROS) and reactive nitrogen species (RNS), serves as a double-edged sword, initiating various diseases if unchecked [3]. Acute and chronic inflammation, two distinct responses to external stimulation like physical and chemical damages, pathogenic microorganisms, and autoimmune reactions, play crucial roles in both host defense and pathological progression [4]. Notably, the occurrence of oxidative stress can activate the signaling pathways related to the inflammatory responses and increase the level of proinflammatory cytokines and molecules, thereby triggering tissue inflammation [5,6].

Conversely, inflammatory stimulation exacerbates the occurrence of oxidative stress by inducing the generation of excessive free radicals, creating a vicious cycle between these two processes. Apoptosis, a highly regulated form of programmed cell death, relies on various signaling networks that converge on a comprehensive action mechanism [7]. The activation of apoptotic pathways is crucial for eliminating damaged cells within tissues and maintaining microenvironmental homeostasis [8]. Moreover, numerous studies have confirmed that the apoptotic signaling pathways can be activated or inhibited by oxidative stress and inflammation, leading to cell death in the end [6,9]. The intricate interdependence among oxidative stress, inflammation, and apoptosis is of paramount importance in livestock and poultry production. In cases where these processes are dysregulated, they contribute significantly to the pathogenesis of numerous disorders, impacting production performance, such as reduced growth rates, lower reproductive efficiency, and compromised meat quality, as well as decreasing disease resistance, thereby enhancing susceptibility to infections and inflammatory conditions [10–12]. Therefore, a comprehensive understanding of the molecular mechanisms underlying these biological processes, along with the identification of potential targets for intervention, is crucial for developing effective strategies to enhance livestock and poultry production performance and maintain physical health.

On account of the intimate link among redox homeostasis, the inflammatory response, and cell apoptotic death in biological systems, increasing research has been conducted on these properties of medicinal homologous foods, particularly their relevance to the production of livestock animals and poultry. These foods prominently embody the value of food and medicine fusion, and their unique biological activities that hold significant potential for enhancing animal health and productivity [13]. The bioactivities of medicinal homologous food extracts are closely related to their chemical compositions, including alkaloids, terpenoids, saponins, and flavonoids, which jointly promote their overall functional effects [14]. Ellagic acid (EA), a polyphenol substance found in natural resources like raspberries and pomegranates, mainly exists in its condensed form [15]. EA exhibits remarkable potential in the treatment of stress-induced pathologies in animals due to its ability to antagonize intracellular oxidative stress, inflammation, and cell apoptosis [16–18]. Meanwhile, the favorable safety profile and compatibility with other therapeutic agents of EA suggest its potential use as a natural additive or supplement in animal feeds, aiming to improve animal resilience and overall performance [19]. Consequently, the exploration of the potential application of EA in addressing oxidative stress-mediated inflammatory injury and cell apoptotic death in livestock and poultry represents a valuable theoretical framework and provides a promising avenue for enhancing animal welfare, health, and productivity. It is worth noting that there have been research results using poultry as model animals, which have found that EA plays an important positive role in improving poultry production performance and resisting harmful stimulation, providing a theoretical basis for its practical application in animal production.

2. The Chemical Properties of Ellagic Acid

Plant polyphenols are secondary metabolites derived from the metabolic pathways of shikimic acid and phenylalanine in plants, containing at least one phenolic hydroxyl group [20]. They can be divided into three categories: tannins, flavonoids, and lignin. EA belongs to the class of polyphenol extractives (tannins) widely spread among dicotyledons. Natural EA mainly exists in concentrated form in the ester linked to sugar in the hydrolyzable tannin component known as ellagitannins (ETs) [15]. During plant chemical processing, under either acidic or basic conditions, ester bonds of ETs are hydrolyzed, yielding a hexahydroxydiphenoyl group, which spontaneously lactonizes into EA [21]. The chemical formula of EA is $C_{14}H_6O_8$, with a molecular weight of 302.19 g/mol, known as 2,3,7,8-tetrahydroxy-chromeno [5,4,3-cde]chromene-5,10-dione. Pure EA is a milky white crystal with a melting point above 360 °C. It is a highly symmetrical molecule, where one molecule of EA is composed of hydrophilic parts of four hydroxyls and two lactones functional groups and lipophilic parts of four six-membered hydrocarbon rings, resulting in the whole molecular space maintaining a planar distribution [22]. Due to the interaction of hydroxyl groups and the lactone systems forming hydrogen bonds, functions such as participating in antioxidant redox reactions are performed by EA [23]. Under the influence of its structure, EA shows poor aqueous solubility, which is the major pharmacological limitation; however, it is slightly soluble in methanol, alcohol, and dimethyl sulfoxide, and the best solvent is polyethylene glycol [24,25].

3. The Anti-Oxidant Property of Ellagic Acid

Oxygen free radicals, especially ROS, are generated intracellularly within the mitochondria and other cellular components [26,27]. Under physiological conditions, low levels of ROS are crucial for maintaining many important physiological functions by regulating redox homeostasis and key signaling pathways. When the ROS accumulates excessively, they can cause the occurrence of oxidative stress and destroy all of the key biomolecules, including proteins, membrane lipids, and DNA, ultimately leading to various diseases such as intestinal inflammation and digestive dysfunction, immunologic suppression, reproductive disorders, and muscle dysplasia and growth retardation in livestock animals and poultry [28–31]. The characteristic of oxidative stress is the transition from the balance between the formation and elimination of free radicals to their excessive formation. To alleviate the adverse effect of ROS, organisms have evolved a sophisticated and multilevel antioxidant defense system, in which the initial and most effective layer comprises enzymes with superior catalytic activity, including glutathione peroxidases (GPx), superoxide dismutases (SODs), thioredoxin, and catalase (CAT), and some antioxidant substrates [32,33]. This intricate system efficiently protects cells from ROS-induced toxicity and alleviates the impairment caused by oxidative stress [34].

Known for its remarkable anti-oxidative stress characteristics, EA stands out as a naturally obtainable pharmacological agent, with a bioactive polyphenolic molecule structure. Numerous studies from both in vivo and in vitro trials have consistently demonstrated the protective function of EA in alleviating oxidative damage in animals caused by various adverse factors, such as heat stress, dietary changes, and hazardous substance exposure [35]. The antioxidant function of EA is achieved by either attenuating the accumulation of excessive ROS or activating specific antioxidant enzymes [36,37]. A previous study showed that dietary supplementation with EA (100, 200, or d 400 mg/kg) could improve serum GPx activity, increase intestine villus height and crypt depth, and decrease the feed/gain ratio in yellow-feathered broilers [38]. Meanwhile, the study result on the potential effects of EA and mesocarp extract of *Punica granatum* on the productive and reproduction performance of laying hens indicated that the supplementation of EA or *Punica granatum* extract to diet remarkably restrained malondialdehyde (MDA) concentration and increased total antioxidant capacity in serum and liver samples, accompanied with increased egg production, feed intake, and body weight of hens [39]. Furthermore, the addition of different doses of EA (0, 75, 150, 300, and 600 mg/kg) to the basic diet can significantly increase the activities of antioxidant enzymes GPx, SOD, and CAT and reduce the level of peroxide MDA in the serum of Arbor Acres broilers under heat stress, thus improving intestinal barrier function and growth performance parameters (Final weight and the feed/gain ratio) [35].

Interestingly, the molecular mechanism by which EA activates the antioxidant system has also been extensively studied and confirmed. It is reported that nuclear factor erythroid 2-related factor 2 (Nrf2) is the key target for EA to stabilize ROS levels [40,41]. As a classical nuclear receptor transcription factor, Nrf2 is a crucial regulator of redox homeostasis. Under quiescent conditions, two glycine repeat domains of kelch-like ECH-associated protein 1 (Keap1) homodimer bind with the latch and the hinge domains of a single Nrf2 molecule to repress Nrf2 activity; however, when cells are exposed to stress resources, the glycine repeat domains of Keap1 released the hinge domain of Nrf2 to block Nrf2 ubiquitination and degradation, thus inducing free and newly synthesized Nrf2 to translocate to the nucleus to achieve its active function [42,43]. Recent studies have demonstrated that

EA can directly stimulate the nuclear translocation of Nrf2, enhancing its interaction with the antioxidant response element sequence found in antioxidant enzymes such as heme oxygenase-1 (HO-1) and quinone oxidoreductase 1 (NQO1) [31,44]. This process increases the expression of these enzymes, which, in turn, helps to alleviate oxidative stress and offers a defensive and protective mechanism against the detrimental effects of xenobiotic metabolism [45,46]. For instance, it has been reported that supplementing with EA in orally administered *Lactobacillus brevis* can significantly increase the mRNA and protein expression of Nrf2, promote the mRNA expression levels of HO-1, NQO1, and GPx1, and increase the contents of SOD, CAT, and glutathione in the caeca of Leghorn chickens [47]. Meanwhile, a recent study has shown that EA has the ability to exert protective effects on tissue dysfunction and structural alteration through the improvement in antioxidant function regulated by sirtuin 1 (SIRT1) [48]. SIRT1 is a histone deacetylase that can sense energy levels in cells and has the function of delaying cell aging, helping cells resist external stress, and improving metabolism. It has been reported that SIRT1 activates Nrf2 nuclear translocation through deacetylation, thereby regulating the expression of its downstream antioxidant and detoxification gene [49]. In addition, besides SIRT1, AMP-activated protein kinase (AMPK) is also targeted and inhibited by EA to alleviate LPS-induced oxidative stress [50]. AMPK can phosphorylate Nrf2 or its upstream signaling molecules, indirectly enhancing the defense ability of the antioxidant system [51]. Therefore, based on existing evidence, activating Nrf2 activity is a key mechanism for EA to exert antioxidant function and maintain tissue cell redox homeostasis (Figure 1). In conclusion, an in-depth understanding and exploration of the regulatory mechanism of EA on redox homeostasis provide substantial help for the subsequent use of EA in preventing adverse consequences related to oxidative stress.

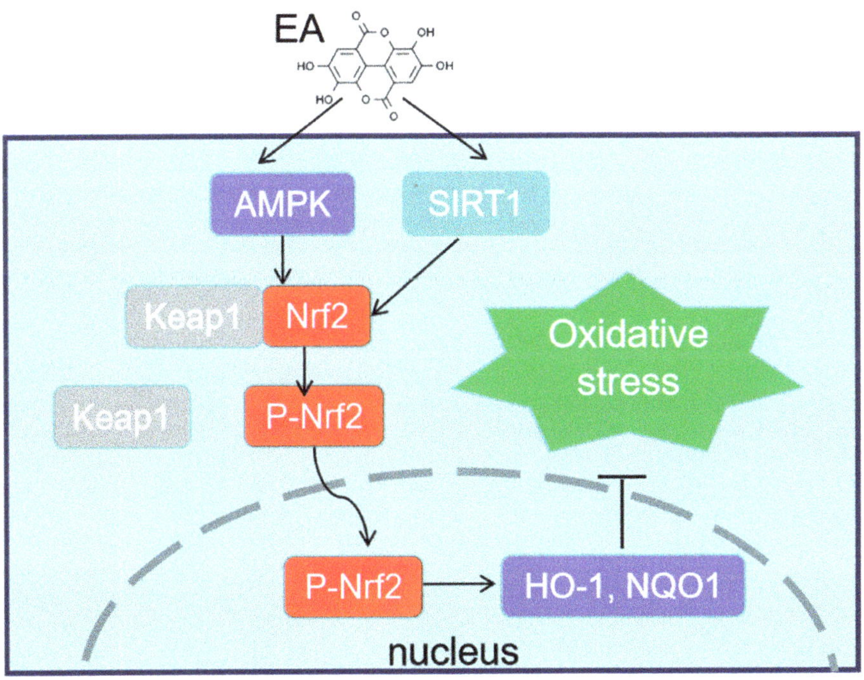

Figure 1. The mechanism of ellagic acid EA in improving the antioxidant system function of poultry. EA = ellagic acid; SRIT = sirtuin 1; AMPK = AMP-activated protein kinase; Keap1 = kelch-like ECH-associated protein 1; Nrf2 = nuclear factor erythroid 2-related factor 2; P-Nrf2 = phosphorylated Nrf2; HO-1 = heme oxygenase-1; NQO1 = quinone oxidoreductase 1.

4. The Anti-Inflammatory Property of Ellagic Acid

Inflammation, a highly conserved and sophisticated defense mechanism, serves as a crucial response to diverse stimulations ranging from microbial invasion, mechanical trauma, or chemical insults, and its primary objective is to neutralize the source of damage and reinstate tissue homeostasis, which is important for maintaining animal health and performance [52]. As a response to these stimuli, immune cells in injured tissues release proinflammatory cytokines, chemokines, and inflammatory metabolites, collectively known as inflammatory mediators [53]. These mediators play an important role in orchestrating the inflammatory response, which is indispensable for combating infections and facilitating tissue repair. However, once the delicate balance between the generation and elimination of inflammatory mediators is disrupted, the tissue will experience sustained or excessive inflammation in livestock and poultry. This dysregulation can develop into different types of diseases, such as intestinal lesions that impair nutrient absorption, metabolic disturbances affecting growth and reproduction, and autoimmune disorders that compromise overall health and welfare [54,55]. Modulating inflammatory mediators emerges as a strategic approach for preventive and therapeutic interventions in livestock and poultry. By targeting specific inflammatory mediators or pathways, researchers aim to mitigate the detrimental consequences of excessive inflammation, enhance animal resilience against disease challenges, and ultimately optimize production efficiency and animal health outcomes.

As a substance endowed with the ability to regulate numerous chemicals and signaling pathways that are involved in the inflammatory cascade, EA exhibits a remarkable potential to modulate the inflammatory response. A series of studies have shown that the reason why EA exerts anti-inflammatory effects is attributed to inhibiting the actions of several mediators that promote inflammation, such as proinflammatory cytokines (interferon γ [IFN-γ], interleukin [IL]-1, 1 IL-6, and IL-17) and chemokines (C-C motif ligand [CCL]-2 and CCL-3), or promote the release of the anti-inflammatory cytokine-like IL-10 [56–58]. For example, dietary EA significantly reduced the content levels of IL-6 and TNF-α and increased the levels of IL-2, IL-10, and immunoglobulins IgG, IgM, and IgA in the cecal tissue of Guangxi yellow feathered broiler chickens infected with *Eimeria tenella* [59]. Of note, the nuclear factor kappa-light-chain-enhancer of activated B cells (NF-κB) pathway, being crucial in the inflammatory process, has been shown to be effectively blocked by EA [60]. NF-κB coordinates the expression of crucial regulatory genes through two distinct activation mechanisms, namely the canonical and noncanonical pathways [61]. The canonical pathway of NF-κB, triggered by inflammatory and immune signals, utilizes the IκB kinase (IKK) complex (IKKγ, IKKα, and IKKβ) to phosphorylate IκB, releasing NF-κB dimers to regulate transcription of proinflammatory cytokines such as tumor necrosis factor-α (TNF-α) and IL-1 [62,63]. EA mainly alleviates inflammation via the NF-κB canonical pathway. Previous studies have demonstrated that by targeting the IKK-NF-κB pathway, EA could lead to a reduction in the expression of these proinflammatory cytokines and retards the subsequent inflammation occurrence [64,65]. It was reported that under post-infection with *Eimeria tenella*, significant anti-inflammatory effects were observed in chickens orally gavaged with EA and in combination with *Lactobacillus*, which was accompanied by NF-κB pathway expression inhibition [47].

Furthermore, some upstream signaling molecules or pathways have also been reported to be regulated by EA, thereby affecting NF-κB pathway-mediated inflammation processes. For instance, in mouse tissue injury models (such as ulcerative colitis, hepatitis, and neuralgia), EA administration can alleviate NF-κB-activated tissue inflammation and pathological damage by inhibiting the phosphorylation activation of the mitogen-activated protein kinase pathway (c-Jun N-terminal kinase, extracellular regulated kinase 1/2, and p38) or the expression of toll-like receptors (TLR2 and TLR4) [66,67]. Significantly, the NF-κB dimer activates target genes by binding to specific DNA binding sites, often requiring the assistance of other transcription factors. For instance, the signal transducers and transcriptional activators (STAT) family can directly control the transcription expression of NF-κB to participate in the inflammation occurrence [68]. The six primary isoforms of the STAT family (STAT1 to STAT6) play significant roles in regulating signal transduction involved in cell proliferation and death, immunity reaction, and metabolism [69]. The ability of EA to inhibit the expression of the STAT transcription factor family is believed to regulate the inflammatory response [70]. Studies have shown that EA can inhibit the activation of Janus kinase (JAK), thereby limiting the recruitment and phosphorylation of STAT protein and alleviating NF-κB-mediated inflammation [62,71]. Tang et al. investigated the therapeutic effects of dietary EA on subclinical necrotic enteritis in broilers and found that EA could alleviate intestinal barrier dysfunction and inflammatory damage in the jejunum, as well as decrease the feed conversion ratios of broilers by modulating the TLR/NF-κB and JAK3/STAT6 signaling pathways [72]. These studies clearly explain the mechanism of EA influencing the NF-κB pathway and its cascade of events (Figure 2), which contributes to emphasizing its potential use as a medication to treat a range of inflammatory conditions in poultry production.

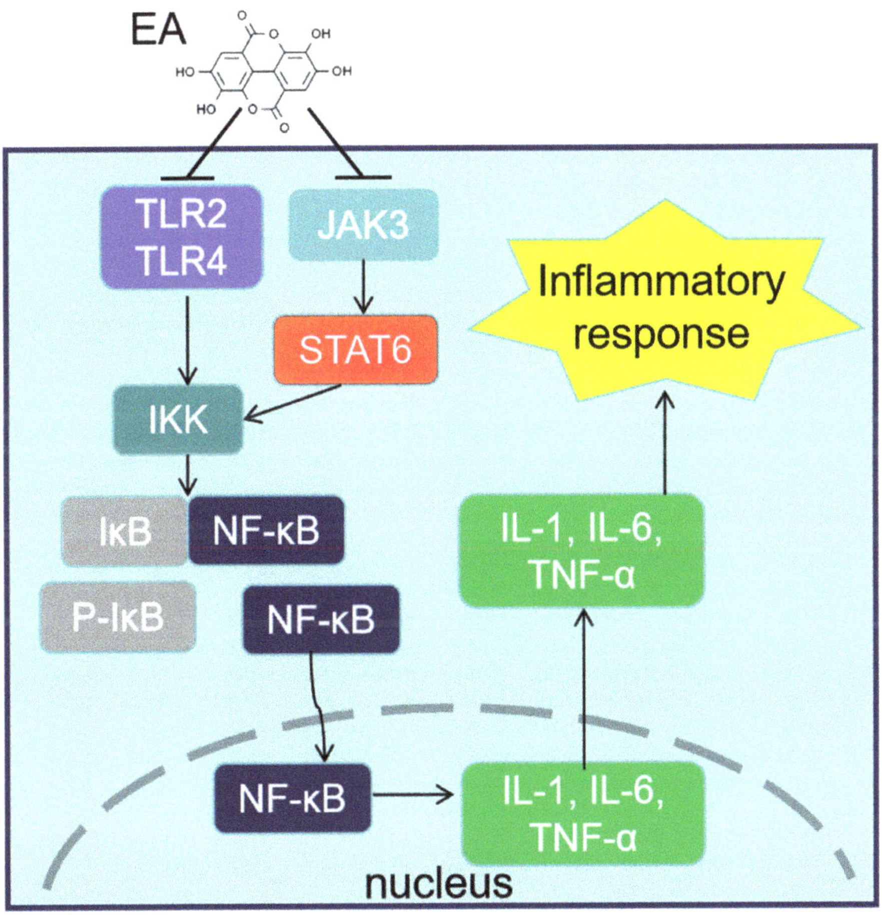

Figure 2. The mechanism of ellagic acid in suppressing the inflammatory response of poultry. EA = ellagic acid; JAK3 = Janus kinase 3; STAT6 = signal transducers and transcriptional activator 6; TLR2 = toll-like receptor 2; TLR4 = toll-like receptor 4; NF-κB = nuclear factor kappa-light-chain-enhancer of activated B cells; IκB = inhibitor of NF-κB; IKK = IκB kinase; P-IκB = phosphorylated IκB; IL-1 = interleukin 1; IL-6 = interleukin 6; TNF-α = tumor necrosis factor-α.

5. The Anti-Apoptosis Property of Ellagic Acid

Apoptosis, universally recognized as a type of programmed cell death, constitutes a crucial biological mechanism essential for maintaining tissue homeostasis by meticulously regulating cell populations. This process selectively eliminates dysfunctional, damaged, or superfluous cells, thereby ensuring the overall health and stability of an organism [73,74]. The occurrence of apoptosis is governed by two primary signaling pathways: the extrinsic and intrinsic pathways [75]. The extrinsic pathway, also known as the death receptor pathway, is initiated when specific ligands bind to transmembrane death receptors (such as Fas) on the cell surface, triggering a cascade of signaling events that ultimately lead to apoptosis [76]. In contrast, the intrinsic pathway is activated in response to intracellular stress signals, such as oxidative stress and inflammation, ultimately leading to permeabilization of the outer mitochondrial membrane and releasing proapoptotic factors that further propagate the apoptotic signaling [7]. The final execution of apoptosis relies on the caspase family of proteases, which are activated by both the extrinsic and intrinsic pathways. These proteolytic enzymes function as molecular scissors, cleaving key cellular substrates to dismantle the cells in a highly energy-dependent manner [77]. However, under the influence of excessive oxidative stress and uncontrolled inflammation, damaged tissues can also undergo excessive apoptosis, which can have a negative impact on the normal physiological functions of livestock and poultry, thereby endangering their production performance [78].

EA can maintain cell viability by inhibiting the activation of pro-apoptotic pathways. In a rat liver injury model induced by methotrexate exposure, oral administration of EA significantly activated the expression of antioxidant genes Nrf2 and HO-1 and inhibited the expression of proinflammatory factors NF-κB and IL-6, thereby alleviating methotrexate-induced mitochondrial dysfunction and apoptosis in liver cells [79]. Moreover, EA treatment significantly downregulated the STAT3/NF-κB pathway and improved testosterone propionate-induced benign prostate hyperplasia in mice by decreasing the inflammatory response, increasing mitochondrial dynamics, and suppressing apoptotic death [62]. Therefore, it can be concluded that EA mainly reduces tissue cell apoptosis caused by pathogenic stimuli in various diseases by exerting its biological effects of antioxidant and anti-inflammatory effects. In the field of poultry production, research on the anti-apoptotic potential of EA is still at a nascent stage. Encouragingly, in a quail model of lead poisoning, the supplementation of EA in diets could significantly alleviate the adverse effects of lead on the growth performance parameters of laying quails (*Coturnix japonica*) and suppress apoptosis in the liver tissue by enhancing the antioxidant defense system [80]. Meanwhile, the results from in vitro cellular experiments have demonstrated that after storing the semen of ROSS 380 roosters in Beltsville extender containing EA liposomes for a certain period of time, compared with the control, EA liposomes significantly increased the GPx and total antioxidant capacity of sperm and reduced the rate of apoptosis, thus accompanying an increase in total motility and membrane functionality [81]. Therefore, we have reason to believe that dietary supplementation with EA has the ability to help poultry reduce the occurrence of abnormal apoptosis of tissue cells. However, whether the anti-apoptotic role of EA in poultry farming to maintain tissue and cellular health is universal remains to be confirmed through extensive practical applications.

6. Conclusions and Perspectives

EA, a bioactive compound present in various plants, performs significant roles in the prevention of diseases, such as immune-related diseases, intestinal injury, and metabolic disorders. This mini-review summarizes an overview of its recent research progress in preventing and treating oxidative stress, inflammation, and cell apoptosis in poultry production, and provides relevant reports on how it helps poultry improve production performance and resist harmful stimuli by regulating these biological processes. Generally, EA mainly exhibits antioxidant properties through Keap1/Nrf2 pathway activation, anti-inflammation properties through NF-κB pathway inhibition, and reduces the incidence of abnormal apoptotic death under the dynamic changes in these two key biological processes (Figure 3). EA has broad application prospects in animal production, including the development of new drugs in the context of antibiotic-free feeding, exploration of alternative treatment methods, clinical trials of targeted therapies for specific diseases, and the combination of feeding interventions to maintain overall health. However, advancing research requires a comprehensive understanding of the mechanisms of EA in reducing diseases and the development of more effective nutritional regulation strategies.

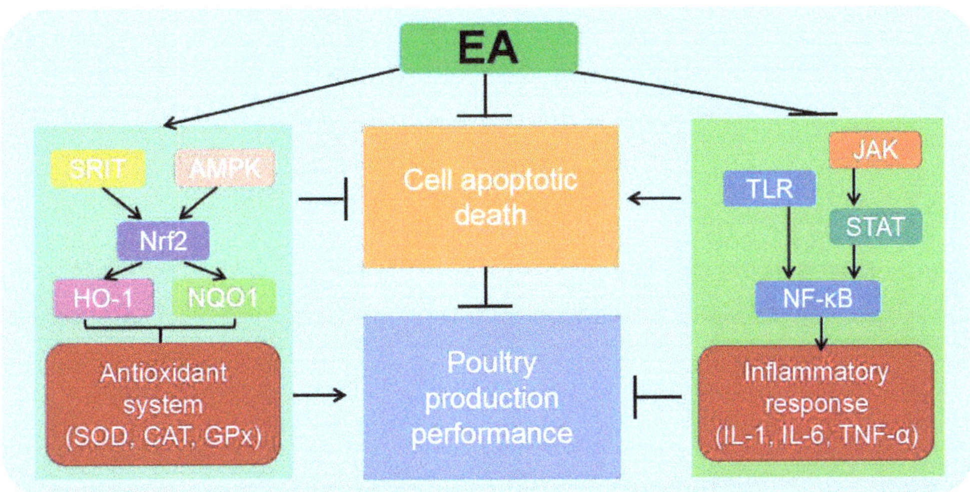

Figure 3. A graphic summary of ellagic acid in improving poultry production performance via regulation of redox homeostasis, the inflammatory response, and cell apoptosis. EA = ellagic acid; SRIT = sirtuin 1; AMPK = AMP-activated protein kinase; Nrf2 = nuclear factor erythroid 2-related factor 2; HO-1 = heme oxygenase-1; NQO1 = quinone oxidoreductase 1; SOD = superoxide dismutases; CAT = catalase; GPx = glutathione peroxidases; JAK = Janus kinase; STAT = signal transducers and transcriptional activators; TLR = toll-like receptor; NF-κB = nuclear factor kappa-light-chain-enhancer of activated B cells; IL-1 = interleukin 1; IL-6 = interleukin 6; TNF-α = tumor necrosis factor-α.

Author Contributions: S.W. and W.Z.: writing—original draft preparation. B.T., Y.H., T.L., X.C. and L.Z.: analysis of data and investigation. X.L.: review and supervision. All authors have read and agreed to the published version of the manuscript.

Funding: This work was funded by the National Key R&D Program of China (2023YFD1301200), the Young Elite Scientists Sponsorship Program by CAST (2022QNRC001), the Jiangsu Shuang Chuang Tuan Dui program (JSSCTD202147), and the Jiangsu Shuang Chuang Ren Cai program (JSSCRC2021541).

Informed Consent Statement: Informed consent was obtained from all subjects involved in the study.

Conflicts of Interest: The authors declare no conflicts of interest.

References

1. Li, R.; Luan, F.; Zhao, Y.; Wu, M.; Lu, Y.; Tao, C.; Zhu, L.; Zhang, C.; Wan, L. Crataegus pinnatifida: A botanical, ethnopharmacological, phytochemical, and pharmacological overview. *J. Ethnopharmacol.* **2023**, *301*, 115819. [CrossRef]
2. Basiouni, S.; Tellez-Isaias, G.; Latorre, J.D.; Graham, B.D.; Petrone-Garcia, V.M.; El-Seedi, H.R.; Yalcin, S.; El-Wahab, A.A.; Visscher, C.; May-Simera, H.L.; et al. Anti-Inflammatory and Antioxidative Phytogenic Substances against Secret Killers in Poultry: Current Status and Prospects. *Vet. Sci.* **2023**, *10*, 55. [CrossRef]
3. Kang, R.; Li, R.; Dai, P.; Li, Z.; Li, Y.; Li, C. Deoxynivalenol induced apoptosis and inflammation of IPEC-J2 cells by promoting ROS production. *Environ. Pollut.* **2019**, *251*, 689–698. [CrossRef]

4. Hybertson, B.M.; Gao, B.; Bose, S.K.; McCord, J.M. Oxidative stress in health and disease: The therapeutic potential of Nrf2 activation. *Mol. Asp. Med.* **2011**, *32*, 234–246. [CrossRef]
5. Arulselvan, P.; Fard, M.T.; Tan, W.S.; Gothai, S.; Fakurazi, S.; Norhaizan, M.E.; Kumar, S.S. Role of Antioxidants and Natural Products in Inflammation. *Oxid. Med. Cell. Longev.* **2016**, *2016*, 5276130. [CrossRef]
6. Hussain, T.; Tan, B.; Yin, Y.; Blachier, F.; Tossou, M.C.; Rahu, N. Oxidative Stress and Inflammation: What Polyphenols Can Do for Us? *Oxid. Med. Cell. Longev.* **2016**, *2016*, 7432797. [CrossRef]
7. Bertheloot, D.; Latz, E.; Franklin, B.S. Necroptosis, pyroptosis and apoptosis: An intricate game of cell death. *Cell. Mol. Immunol.* **2021**, *18*, 1106–1121. [CrossRef]
8. Nagata, S. Apoptosis and autoimmune diseases. *Ann. N. Y. Acad. Sci.* **2010**, *1209*, 10–16. [CrossRef]
9. Chen, Z.; Wang, C.; Yu, N.; Si, L.; Zhu, L.; Zeng, A.; Liu, Z.; Wang, X. INF2 regulates oxidative stress-induced apoptosis in epidermal HaCaT cells by modulating the HIF1 signaling pathway. *Biomed. Pharmacother.* **2019**, *111*, 151–161. [CrossRef]
10. Hao, Y.; Xing, M.; Gu, X. Research Progress on Oxidative Stress and Its Nutritional Regulation Strategies in Pigs. *Animals* **2021**, *11*, 1384. [CrossRef]
11. Broom, L.J.; Kogut, M.H. Inflammation: Friend or foe for animal production? *Poult. Sci.* **2018**, *97*, 510–514. [CrossRef]
12. Wang, H.; Wang, A.; Wang, X.; Zeng, X.; Xing, H. AMPK/PPAR-γ/NF-κB axis participates in ROS-mediated apoptosis and autophagy caused by cadmium in pig liver. *Environ. Pollut.* **2022**, *294*, 118659. [CrossRef]
13. Hasted, T.L.; Sharif, S.; Boerlin, P.; Diarra, M.S. Immunostimulatory Potential of Fruits and Their Extracts in Poultry. *Front. Immunol.* **2021**, *12*, 641696. [CrossRef]
14. Zheng, J.; Shang, M.; Dai, G.; Dong, J.; Wang, Y.; Duan, B. Bioactive polysaccharides from Momordica charantia as functional ingredients: A review of their extraction, bioactivities, structural-activity relationships, and application prospects. *Crit. Rev. Food Sci. Nutr.* **2023**, *2023*, 1–24. [CrossRef]
15. Zhang, M.; Cui, S.; Mao, B.; Zhang, Q.; Zhao, J.; Zhang, H.; Tang, X.; Chen, W. Ellagic acid and intestinal microflora metabolite urolithin A: A review on its sources, metabolic distribution, health benefits, and biotransformation. *Crit. Rev. Food Sci. Nutr.* **2023**, *63*, 6900–6922. [CrossRef]
16. Alfei, S.; Turrini, F.; Catena, S.; Zunin, P.; Grilli, M.; Pittaluga, A.M.; Boggia, R. Ellagic acid a multi-target bioactive compound for drug discovery in CNS? A narrative review. *Eur. J. Med. Chem.* **2019**, *183*, 111724. [CrossRef]
17. da Silva, F.B.; Romero, W.G.; Rouver, W.D.N.; Silva, K.; de Almeida, S.A.; Mengal, V.; Peluso, A.A.; Endlich, P.W.; Bissoli, N.S.; Claudio, E.R.G.; et al. Ellagic Acid prevents vascular dysfunction in small mesenteric arteries of ovariectomized hypertensive rats. *J. Nutr. Biochem.* **2022**, *105*, 108995. [CrossRef]
18. Xue, P.; Zhang, G.; Zhang, J.; Ren, L. Synergism of ellagic acid in combination with radiotherapy and chemotherapy for cancer treatment. *Phytomedicine* **2022**, *99*, 153998. [CrossRef]
19. Altindag, F.; Ragbetli, M.C.; Ozdek, U.; Koyun, N.; Ismael Alhalboosi, J.K.; Elasan, S. Combined treatment of sinapic acid and ellagic acid attenuates hyperglycemia in streptozotocin-induced diabetic rats. *Food Chem. Toxicol.* **2021**, *156*, 112443. [CrossRef]
20. Iqbal, I.; Wilairatana, P.; Saqib, F.; Nasir, B.; Wahid, M.; Latif, M.F.; Iqbal, A.; Naz, R.; Mubarak, M.S. Plant Polyphenols and Their Potential Benefits on Cardiovascular Health: A Review. *Molecules* **2023**, *28*, 6403. [CrossRef]
21. Evtyugin, D.D.; Magina, S.; Evtuguin, D.V. Recent Advances in the Production and Applications of Ellagic Acid and Its Derivatives. A Review. *Molecules* **2020**, *25*, 2745. [CrossRef]
22. Tokutomi, H.; Takeda, T.; Hoshino, N.; Akutagawa, T. Molecular Structure of the Photo-Oxidation Product of Ellagic Acid in Solution. *ACS Omega* **2018**, *3*, 11179–11183. [CrossRef]
23. Nyamba, I.; Lechanteur, A.; Semde, R.; Evrard, B. Physical formulation approaches for improving aqueous solubility and bioavailability of ellagic acid: A review. *Eur. J. Pharm. Biopharm.* **2021**, *159*, 198–210. [CrossRef]
24. Bala, I.; Bhardwaj, V.; Hariharan, S.; Kumar, M.N. Analytical methods for assay of ellagic acid and its solubility studies. *J. Pharm. Biomed. Anal.* **2006**, *40*, 206–210. [CrossRef]
25. Li, Y.; Zhang, Y.; Dai, W.; Zhang, Q. Enhanced oral absorption and anti-inflammatory activity of ellagic acid via a novel type of case in nanosheets constructed by simple coacervation. *Int. J. Pharm.* **2021**, *594*, 120131. [CrossRef]
26. Jomova, K.; Raptova, R.; Alomar, S.Y.; Alwasel, S.H.; Nepovimova, E.; Kuca, K.; Valko, M. Reactive oxygen species, toxicity, oxidative stress, and antioxidants: Chronic diseases and aging. *Arch. Toxicol.* **2023**, *97*, 2499–2574. [CrossRef]
27. Nolfi-Donegan, D.; Braganza, A.; Shiva, S. Mitochondrial electron transport chain: Oxidative phosphorylation, oxidant production, and methods of measurement. *Redox Biol.* **2020**, *37*, 101674. [CrossRef]
28. Zheng, S.; Jin, X.; Chen, M.; Shi, Q.; Zhang, H.; Xu, S. Hydrogen sulfide exposure induces jejunum injury via CYP450s/ROS pathway in broilers. *Chemosphere* **2019**, *214*, 25–34. [CrossRef]
29. Sun, Y.; Huang, K.; Long, M.; Yang, S.; Zhang, Y. An update on immunotoxicity and mechanisms of action of six environmental mycotoxins. *Food Chem. Toxicol.* **2022**, *163*, 112895. [CrossRef]
30. Xing, T.; Zhao, X.; Wang, P.; Chen, H.; Xu, X.; Zhou, G. Different oxidative status and expression of calcium channel components in stress-induced dysfunctional chicken muscle. *J. Anim. Sci.* **2017**, *95*, 1565–1573. [CrossRef]
31. da Silva, A.S.; Souza, C.F.; Baldissera, M.D.; Von Laer, A.E.; Lovato, L.T.; Sarturi, J.A.; Herrmann, G.P.; de Moura, A.B.; Favaretto, J.A.; Frias-De-Diego, A.; et al. Relation of reproductive disturbance in sheep and Leptospira interrogans serovar Icterohaemorrhagiae infection: Impacts on cellular oxidation status. *Microb. Pathog.* **2019**, *130*, 65–70. [CrossRef]

32. Barati, E.; Nikzad, H.; Karimian, M. Oxidative stress and male infertility: Current knowledge of pathophysiology and role of antioxidant therapy in disease management. *Cell. Mol. Life Sci.* **2020**, *77*, 93–113. [CrossRef]
33. Sies, H. Oxidative stress: A concept in redox biology and medicine. *Redox Biol.* **2015**, *4*, 180–183. [CrossRef]
34. Jomova, K.; Alomar, S.Y.; Alwasel, S.H.; Nepovimova, E.; Kuca, K.; Valko, M. Several lines of antioxidant defense against oxidative stress: Antioxidant enzymes, nanomaterials with multiple enzyme-mimicking activities, and low-molecular-weight antioxidants. *Arch. Toxicol.* **2024**, *98*, 1323–1367. [CrossRef]
35. Yang, T.; Liu, B.; Wang, Y.; Huang, X.; Yan, Z.; Jiang, Q.; Chen, Q. Ellagic Acid Improves Antioxidant Capacity and Intestinal Barrier Function of Heat-Stressed Broilers via Regulating Gut Microbiota. *Animals* **2022**, *12*, 1180. [CrossRef]
36. Aishwarya, V.; Solaipriya, S.; Sivaramakrishnan, V. Role of ellagic acid for the prevention and treatment of liver diseases. *Phytother. Res.* **2021**, *35*, 2925–2944. [CrossRef]
37. Zeb, A. Ellagic acid in suppressing in vivo and in vitro oxidative stresses. *Mol. Cell. Biochem.* **2018**, *448*, 27–41. [CrossRef]
38. Wang, F.; Chen, J.; Yin, Y.; Yang, M.; Xiao, Y.; Cheng, Y.; Yin, L.; Fu, C. The effects of dietary ellagic acid supplementation on growth performance, immune response, antioxidant activity, digestive enzyme activities, and intestinal functions in yellow-feathered broilers. *J. Anim. Sci.* **2022**, *100*, skac301. [CrossRef]
39. Gul, H.; Geng, Z.; Habib, G.; Hayat, A.; Rehman, M.U.; Khan, I. Effect of ellagic acid and mesocarp extract of Punica granatum on productive and reproductive performances of laying hens. *Trop. Anim. Health Prod.* **2022**, *54*, 228. [CrossRef]
40. Keleku-Lukwete, N.; Suzuki, M.; Yamamoto, M. An Overview of the Advantages of KEAP1-NRF2 System Activation during Inflammatory Disease Treatment. *Antioxid. Redox Signal.* **2018**, *29*, 1746–1755. [CrossRef]
41. Yang, X.; Chu, F.; Jiao, Z.; Yu, H.; Yang, W.; Li, Y.; Lu, C.; Ma, H.; Wang, S.; Liu, Z.; et al. Ellagic acid ameliorates arsenic-induced neuronal ferroptosis and cognitive impairment via Nrf2/GPX4 signaling pathway. *Ecotoxicol. Environ. Saf.* **2024**, *283*, 116833. [CrossRef]
42. Dodson, M.; de la Vega, M.R.; Cholanians, A.B.; Schmidlin, C.J.; Chapman, E.; Zhang, D.D. Modulating NRF2 in Disease: Timing Is Everything. *Annu. Rev. Pharmacol. Toxicol.* **2019**, *59*, 555–575. [CrossRef]
43. Yamamoto, M.; Kensler, T.W.; Motohashi, H. The KEAP1-NRF2 System: A Thiol-Based Sensor-Effector Apparatus for Maintaining Redox Homeostasis. *Physiol. Rev.* **2018**, *98*, 1169–1203. [CrossRef]
44. Lee, H.B.; Lee, S.E.; Park, M.J.; Han, D.H.; Lim, E.S.; Ryu, B.; Kim, E.Y.; Park, S.P. Ellagic acid treatment during in vitro maturation of porcine oocytes improves development competence after parthenogenetic activation and somatic cell nuclear transfer. *Theriogenology* **2024**, *215*, 214–223. [CrossRef]
45. Ross, D.; Siegel, D. The diverse functionality of NQO1 and its roles in redox control. *Redox Biol.* **2021**, *41*, 101950. [CrossRef]
46. Shi, J.; Yu, T.; Song, K.; Du, S.; He, S.; Hu, X.; Li, X.; Li, H.; Dong, S.; Zhang, Y.; et al. Dexmedetomidine ameliorates endotoxin-induced acute lung injury in vivo and in vitro by preserving mitochondrial dynamic equilibrium through the HIF-1a/HO-1 signaling pathway. *Redox Biol.* **2021**, *41*, 101954. [CrossRef]
47. Yang, X.; Pan, X.; Jia, Z.; Bai, B.; Zhi, W.; Chen, H.; Ma, C.; Ma, D. Oral administration of *Lactobacillus brevis* 23017 combined with ellagic acid attenuates intestinal inflammatory injury caused by *Eimeria* infection by activating the Nrf2/HO-1 antioxidant pathway. *Vet Res* **2022**, *53*, 21. [CrossRef]
48. Naghibi, N.; Sadeghi, A.; Movahedinia, S.; Rahimi Naiini, M.; Rajizadeh, M.A.; Bahri, F.; Nazari-Robati, M. Ellagic acid ameliorates aging-induced renal oxidative damage through upregulating SIRT1 and NRF2. *BMC Complement. Med. Ther.* **2023**, *23*, 77. [CrossRef]
49. Li, S.; Jiang, X.; Luo, Y.; Zhou, B.; Shi, M.; Liu, F.; Sha, A. Sodium/calcium overload and Sirt1/Nrf2/OH-1 pathway are critical events in mercuric chloride-induced nephrotoxicity. *Chemosphere* **2019**, *234*, 579–588. [CrossRef]
50. Tanaka, M.; Kishimoto, Y.; Sasaki, M.; Sato, A.; Kamiya, T.; Kondo, K.; Iida, K. *Terminalia bellirica* (Gaertn.) Roxb. Extract and Gallic Acid Attenuate LPS-Induced Inflammation and Oxidative Stress via MAPK/NF-κB and Akt/AMPK/Nrf2 Pathways. *Oxidative Med. Cell. Longev.* **2018**, *2018*, 9364364. [CrossRef]
51. Cui, Y.; Liu, Q.; Zhang, Q.; Di, X.; Zhang, H. Benzoylaconine Protects Skeletal Muscle Against Ischemia-Reperfusion Injury Through Activation of IF1-Dependent AMPK/Nrf2 Axis. *Drug Des. Dev. Ther.* **2024**, *18*, 2125–2142. [CrossRef]
52. Leuti, A.; Fazio, D.; Fava, M.; Piccoli, A.; Oddi, S.; Maccarrone, M. Bioactive lipids, inflammation and chronic diseases. *Adv. Drug Deliv. Rev.* **2020**, *159*, 133–169. [CrossRef]
53. Zhao, H.; Wu, L.; Yan, G.; Chen, Y.; Zhou, M.; Wu, Y.; Li, Y. Inflammation and tumor progression: Signaling pathways and targeted intervention. *Signal Transduct. Target. Ther.* **2021**, *6*, 263. [CrossRef]
54. Cao, Z.; Mu, S.; Wang, M.; Zhang, Y.; Zou, G.; Yuan, X.; Huang, Y.; Yu, S.; Zhang, J.; Zhang, C. Succinate pretreatment attenuates intestinal ischemia-reperfusion injury by inhibiting necroptosis and inflammation via upregulating Klf4. *Int. Immunopharmacol.* **2023**, *120*, 110425. [CrossRef]
55. Mahajan, H.; Lieber, J.; Carson Mallinson, P.A.; Bhogadi, S.; Banjara, S.K.; Kinra, S.; Kulkarni, B. The higher dietary inflammation is associated with a higher burden of multimorbidity of cardio-metabolic and mental health disorders in an urbanizing community of southern India: A cross-sectional analysis for the APCAPS cohort. *Hum. Nutr. Metab.* **2024**, *36*, 200254. [CrossRef]
56. Arab, H.H.; Gad, A.M.; Fikry, E.M.; Eid, A.H. Ellagic acid attenuates testicular disruption in rheumatoid arthritis via targeting inflammatory signals, oxidative perturbations and apoptosis. *Life Sci.* **2019**, *239*, 117012. [CrossRef]

57. Jafari Karegar, S.; Aryaeian, N.; Hajiluian, G.; Suzuki, K.; Shidfar, F.; Salehi, M.; Ashtiani, B.H.; Farhangnia, P.; Delbandi, A.A. Ellagic acid effects on disease severity, levels of cytokines and T-bet, RORγt, and GATA3 genes expression in multiple sclerosis patients: A multicentral-triple blind randomized clinical trial. *Front. Nutr.* **2023**, *10*, 1238846. [CrossRef]
58. Sanadgol, N.; Golab, F.; Mostafaie, A.; Mehdizadeh, M.; Abdollahi, M.; Sharifzadeh, M.; Ravan, H. Ellagic acid ameliorates cuprizone-induced acute CNS inflammation via restriction of microgliosis and down-regulation of CCL2 and CCL3 pro-inflammatory chemokines. *Cell Mol. Biol. (Noisy-Le-Grand)* **2016**, *62*, 24–30.
59. Yang, Y.; Ding, X.; Memon, F.U.; Zhang, G.; Jiang, M.; Hu, D.; Si, H. Punicalagin: A monomer with anti-Eimeria tenella effect from fruit peel of *Punica granatum* L. *Poult. Sci.* **2022**, *101*, 102100. [CrossRef]
60. Garcia, G.; Pais, T.F.; Pinto, P.; Dobson, G.; McDougall, G.J.; Stewart, D.; Santos, C.N. Bioaccessible Raspberry Extracts Enriched in Ellagitannins and Ellagic Acid Derivatives Have Anti-Neuroinflammatory Properties. *Antioxidants* **2020**, *9*, 970. [CrossRef]
61. Zinatizadeh, M.R.; Schock, B.; Chalbatani, G.M.; Zarandi, P.K.; Jalali, S.A.; Miri, S.R. The Nuclear Factor κ B (NF-kB) signaling in cancer development and immune diseases. *Genes Dis.* **2021**, *8*, 287–297. [CrossRef]
62. Park, W.Y.; Song, G.; Park, J.Y.; Ahn, K.S.; Kwak, H.J.; Park, J.; Lee, J.H.; Um, J.Y. Ellagic acid improves benign prostate hyperplasia by regulating androgen signaling and STAT3. *Cell Death Dis.* **2022**, *13*, 554. [CrossRef]
63. Yu, H.; Lin, L.; Zhang, Z.; Zhang, H.; Hu, H. Targeting NF-κB pathway for the therapy of diseases: Mechanism and clinical study. *Signal Transduct. Target. Ther.* **2020**, *5*, 209. [CrossRef]
64. Bains, M.; Kaur, J.; Akhtar, A.; Kuhad, A.; Sah, S.P. Anti-inflammatory effects of ellagic acid and vanillic acid against quinolinic acid-induced rat model of Huntington's disease by targeting IKK-NF-κB pathway. *Eur. J. Pharmacol.* **2022**, *934*, 175316. [CrossRef]
65. Sun, S.C. The non-canonical NF-κB pathway in immunity and inflammation. *Nat. Rev. Immunol.* **2017**, *17*, 545–558. [CrossRef]
66. Lee, J.H.; Won, J.H.; Choi, J.M.; Cha, H.H.; Jang, Y.J.; Park, S.; Kim, H.G.; Kim, H.C.; Kim, D.K. Protective effect of ellagic acid on concanavalin A-induced hepatitis via toll-like receptor and mitogen-activated protein kinase/nuclear factor κB signaling pathways. *J. Agric. Food Chem.* **2014**, *62*, 10110–10117. [CrossRef]
67. Marin, M.; Maria Giner, R.; Rios, J.L.; Recio, M.C. Intestinal anti-inflammatory activity of ellagic acid in the acute and chronic dextrane sulfate sodium models of mice colitis. *J. Ethnopharmacol.* **2013**, *150*, 925–934. [CrossRef]
68. Mao, P.; Wang, T.; Du, C.W.; Yu, X.; Wang, M.D. CXCL5 promotes tumorigenesis and angiogenesis of glioblastoma via JAK-STAT/NF-κb signaling pathways. *Mol. Biol. Rep.* **2023**, *50*, 8015–8023. [CrossRef]
69. Wong, G.L.; Manore, S.G.; Doheny, D.L.; Lo, H.W. STAT family of transcription factors in breast cancer: Pathogenesis and therapeutic opportunities and challenges. *Semin. Cancer Biol.* **2022**, *86*, 84–106. [CrossRef]
70. Liu, Q.; Liang, X.; Liang, M.; Qin, R.; Qin, F.; Wang, X. Ellagic Acid Ameliorates Renal Ischemic-Reperfusion Injury through NOX4/JAK/STAT Signaling Pathway. *Inflammation* **2020**, *43*, 298–309. [CrossRef]
71. Huang, P.Y.; Hsieh, Y.H.; Ting, Y.H.; Lee, C.C.; Tsai, J.P. Ellagic acid ameliorates renal fibrogenesis by blocking epithelial-to-mesenchymal transition. *Tzu Chi Med. J.* **2024**, *36*, 59–66. [CrossRef]
72. Tang, Y.; Zhang, X.; Wang, Y.; Guo, Y.; Zhu, P.; Li, G.; Zhang, J.; Ma, Q.; Zhao, L. Dietary ellagic acid ameliorated Clostridium perfringens-induced subclinical necrotic enteritis in broilers via regulating inflammation and cecal microbiota. *J. Anim. Sci. Biotechnol.* **2022**, *13*, 47. [CrossRef]
73. Pritchard, D.M.; Watson, A.J. Apoptosis and gastrointestinal pharmacology. *Pharmacol. Ther.* **1996**, *72*, 149–169. [CrossRef]
74. Tower, J. Programmed cell death in aging. *Ageing Res. Rev.* **2015**, *23*, 90–100. [CrossRef]
75. Pistritto, G.; Trisciuoglio, D.; Ceci, C.; Garufi, A.; D'Orazi, G. Apoptosis as anticancer mechanism: Function and dysfunction of its modulators and targeted therapeutic strategies. *Aging* **2016**, *8*, 603–619. [CrossRef]
76. Kashyap, D.; Garg, V.K.; Goel, N. Intrinsic and extrinsic pathways of apoptosis: Role in cancer development and prognosis. *Adv. Protein Chem. Struct. Biol.* **2021**, *125*, 73–120. [CrossRef]
77. Lossi, L. The concept of intrinsic versus extrinsic apoptosis. *Biochem. J.* **2022**, *479*, 357–384. [CrossRef]
78. Dong, W.; Liu, M.; Liu, B.; Xiao, Y.; Liu, X.; Yang, M.; Yuan, X.; Zhang, Y.; Li, G.; Meng, K. Isolation of Bacillus licheniformis and its protective effect on liver oxidative stress and apoptosis induced by aflatoxin B1. *Poult. Sci.* **2024**, *103*, 104079. [CrossRef]
79. Ebrahimi, R.; Sepand, M.R.; Seyednejad, S.A.; Omidi, A.; Akbariani, M.; Gholami, M.; Sabzevari, O. Ellagic acid reduces methotrexate-induced apoptosis and mitochondrial dysfunction via up-regulating Nrf2 expression and inhibiting the IκBα/NFκB in rats. *DARU J. Pharm. Sci.* **2019**, *27*, 721–733. [CrossRef]
80. Iflazoglu Mutlu, S.; Seven, I.; Arkali, G.; Birben, N.; Sur Arslan, A.; Aksakal, M.; Tatli Seven, P. Ellagic acid plays an important role in enhancing productive performance and alleviating oxidative stress, apoptosis in laying quail exposed to lead toxicity. *Ecotoxicol. Environ. Saf.* **2021**, *208*, 111608. [CrossRef]
81. Najafi, A.; Taheri, R.A.; Mehdipour, M.; Martinez-Pastor, F.; Rouhollahi, A.A.; Nourani, M.R. Improvement of post-thawed sperm quality in broiler breeder roosters by ellagic acid-loaded liposomes. *Poult. Sci.* **2019**, *98*, 440–446. [CrossRef]

animals

MDPI

Article

Tea Polyphenol Protects the Immune Barrier and Inhibits TLR2/NF-κB/MLCK Signal Activation to Prevent Inflammatory Injury in the Intestines of Common Carp (*Cyprinus carpio* L.)

Man Qian, Jie Yang, Yao Xue, Jiawei Wu, Ziyi Li, Jilong Luo, Bing Zhao * and Xuejiao Gao *

College of Veterinary Medicine, Northeast Agricultural University, Harbin 150030, China; qm18437916979@163.com (M.Q.); yang11jie2022@126.com (J.Y.); 15561850067@163.com (Y.X.); jiayou0609@163.com (J.W.); s230601010@neau.edu.cn (Z.L.)

* Correspondence: z12zhaobing@163.com (B.Z.); xuejiaogao@126.com (X.G.)

Simple Summary: Tea polyphenol (TP) has increasingly attracted wide, attractive attention due to their powerful biological activities. In this study, we found that TP activated intestinal immune ability and, combined with TLR2 to inhibit NF-κB activation, reduced the expression of pro-inflammatory factors. TP also reserved the phosphorylation of MLC to upregulate the expression of tight junction proteins. These results showed that dietary TP protected the intestinal immune barrier and alleviated inflammatory injury through the TLR2/NF-κB/MLCK signal in the intestines of common carp.

Abstract: Tea polyphenol (TP) is a kind of natural macromolecular compound present in tea extract with rich biological potential. The purpose of this study was to explore the protective effect of TP on the intestinal immune barrier and the related mechanisms of TP alleviating intestinal injury. Models of common carp (*Cyprinus carpio* L.) and primary intestinal epithelial cells treated with TP and lipopolysaccharide (LPS) were established. The qPCR and ELISA results showed that TP increased the levels of lysozyme (LZ), alkaline phosphatase (ALP), complement component 3 (C3), complement component 4 (C4), immunoglobulin T (IgT), immunoglobulin D (IgD), and immunoglobulin M (IgM) to activate intestinal immune ability. Molecular docking indicated that TP had a strong interaction with TLR2. Meanwhile, TP alleviated LPS-induced intestinal inflammatory damage as evidenced by reducing the mRNA levels of *TNF-α*, *IL-6*, *IL-1β*, *TLR2*, *MyD88*, *P65*, and *IκBα*, which were consistent with those of ELISA and Western blotting results. Moreover, the qPCR and Western blotting results revealed TP promoted the levels of tight junction-related proteins (claudins, occludin, and ZOs) and inhibited the phosphorylation of MLC, which showed the opposite trend after LPS treatment. In summary, the present study indicated that TP improved immune ability and inhibited the activation of the TLR2/NF-κB/MLCK pathway to attenuate LPS-induced inflammatory injury in the intestines of common carp.

Keywords: tea polyphenol; intestinal immunity; inflammatory injury; TLR2/NF-κB/MLCK pathway

1. Introduction

Common carp (*Cyprinus carpio* L.), one of the most consumed in Asia, occupies an important position in freshwater aquaculture. The intestines are an essential immune organ of fish [1]. Intestinal immunity depends on innate and adaptive immune responses. The innate immune components of the intestine include alkaline phosphatase (ALP), lysozyme (LZ), and complements (such as C3 and C4) in fish. LZ exhibited antibacterial effects by destroying bacterial cell walls to lyse bacteria [2]. Complement, equally as an important antimicrobial component, can directly participate in body defense [3]. Igs are the main adaptive immune components of the intestine. Igs (IgM, IgD, and IgT) act as a local immune system to directly combine with the antigen, preventing pathogens from invading the cell in teleosts [4]. Previous studies have shown that LPS injections can increase complement activity to stimulate immune response and induce inflammatory response in carp [5]. In this study, LPS challenge was used to investigate its effect on intestinal immune level.

Toll-like receptors (TLR) are important protein molecules involved in innate immunity, which are also a bridge connecting innate immunity and adaptive immunity. In common carp, TLR2 is a major protein molecule that participates in the inflammatory signal transduction process [6]. The activation of TLR2 is dependent on the combination with the downstream protein myeloid differentiation factor 88 (MyD88) [7]. Nuclear factor kappa-B (NF-κB) is the hub of the downstream signaling pathway of TLR2. The expression of inflammatory factors is regulated by activated NF-κB, which initiates the immune response [8]. However, excessive inflammatory factors are the key triggers of intestinal mechanical barrier destruction. Tight junction proteins are essential components of the mechanical barrier of the intestine. They close the gaps between cells to prevent harmful substances from leaking into the body [9]. It is now understood that tight junction proteins are regulated by myosin light chain kinase (MLCK). MLCK promotes myosin light chain (MLC) phosphorylation and affects the expression of tight junction proteins, resulting in intestinal barrier dysfunction [10]. In addition, the activation of MLCK is regulated by NF-κB, which affects tight junctions, causing increased intestinal permeability [11].

Tea polyphenol (TP), the main active ingredient in tea, is widely used as food additives [12]. TP has been proven to have a variety of biological activities, including anti-inflammatory, antioxidant, and immune protection effects [13,14]. Studies have pointed out that dietary TP can improve the growth performance of groupers under a high-fat diet [15]. Studies have shown that TP can also alleviate tissue damage caused by poisons such as tetrabromobisphenol A, trimethyltin chloride, and acetochlor [12,16,17]. Although studies have pointed out the protective effect of TP on fish, the intestinal immune capacity and tissue protection of TP still need to be further explored in common carp. This study attempted to explore the beneficial effect of TP on the intestines by detecting histopathological changes, immune factor levels, inflammation, and tight junction indicators in the intestines of common carp, providing a theoretical basis for the application of TP in aquaculture.

2. Materials and Methods

All the experimental procedures were followed by the Institutional Animal Care and Use Committee of Northeast Agriculture University in this study (NEAU2023046).

2.1. Molecular Docking

Molecular docking was used to evaluate the interaction between TP and TLR2. AutoDock (v4.2.6) (http://autodock.scripps.edu/) is the automatic docking procedure for this docking study. The GenBank login number of TLR2 is BAU98381. Pymol (v3.1) software was used to optimize the target, such as removing water molecules and small molecular ligands. Hydrogenation and charge treatment were performed with AutoDock Tools and saved in a "pdbqt" format. The binding energy between TP and TLR2 was calculated according to AutoDock's interaction energy protocol. A binding energy of less than −5.0 kcal/mol represents an ability to interact, and a binding energy of less than −7.0 kcal/mol represents a strong binding activity.

2.2. Animals and Treatments

Eighty common carps (mean initial weight: 152.63 ± 11.52 g) were randomly divided into eight tanks (70 × 50 × 30 cm) and fed on a pellet diet at 3% of their body weight two times every day. Continuous 24 h aeration was supplied to the tanks by air-stones. One-third of the water was replaced with clean dechlorinated water and adaptive feeding for 7 days. After 7 days of acclimation, the fish were randomly distributed into four tanks (20 fish/tank). In the control group (CG), the common carp were fed with a control diet for 21 d and intraperitoneal injection with PBS for 96 h. In the TP group (TP), the common carp were fed with a diet containing 1000 mg/kg TP for 21 d and intraperitoneal injection with PBS for 96 h. In the LPS group (LPS), the common carp were fed with a control diet for 21 d and intraperitoneal injection with 4 mg/kg LPS for 96 h. In the TP + LPS group (TL), the common carp were fed with a diet containing 1000 mg/kg TP for 21 d and intraperitoneally injected with 4 mg/kg LPS for 96 h. The concentrations of LPS and TP were determined based on previous studies [16,18]. Other environmental conditions were consistent with natural conditions. Subsequently, the common carp were anesthetized with 0.02% MS-222 and euthanized. The intestines were removed, a part of the intestines rinsed with PBS, and fixed in 4% paraformaldehyde. The remaining intestinal segments were frozen in liquid nitrogen and stored at −80 °C for later tests.

2.3. Cell Culture

Primary intestinal epithelial cells of common carp were cultured according to the previous study [19]. Common carp were starved, and intestinal segments were aseptically removed. The intestinal segments were rinsed with PBS containing penicillin (200 μg/mL) and streptomycin (200 μg/mL). Then, intestinal segments were cut into pieces and washed with PBS as described above. The intestinal fragments were digested with trypsin solution for 15 min in a 37 °C water bath. After termination of digestion, the mixture was filtrated through a 100 μm strainer and centrifuged. The cell pellet was resuspended in M199 culture medium containing 10% fetal bovine serum, 100 μg/mL penicillin, and streptomycin and cultured in 6-well plates at 27 °C in a 5% CO_2 atmosphere.

The cell viability was detected with a Cell Counting Kit-8 (Biosharp, Hefei, China). Intestinal epithelial cells (1×10^4) were seeded in 96-well plates and treated with TP (10, 20, 40, 80, 160, 320 μg/mL) for 48 h. The CG received 0.1% DMSO (v/v). Furthermore, 10 μL of Cell Counting Kit-8 solution was added to each hole and incubated at 37 °C for 1 h. The absorbance was detected at a wavelength of 450 nm. Relative cell viability was displayed as an absorbance value relative to the control. The viability of primary intestinal epithelial cells could reach more than 90% after being treated with 160 μg/mL TP (Figure S1). Then, 160 μg/mL TP was used for subsequent experiments. LPS (30 μg/mL) concentrations were determined according to previous studies [20].

2.4. Histopathology Observation

The intestinal tissue was fixed with 4% glutaraldehyde and then removed, dehydrated in graded ethanol, permeabilized in xylene, embedded in paraffin, and sectioned into

5-micrometer-thick slices. The slices were stained with H&E and observed with a microscope (Olympus, Tokyo, Japan).

2.5. Real-Time Quantitative PCR

The total RNA of intestinal tissue and cells were extracted by the Trizol method. A UV spectrophotometer was used to measure the concentration of total RNA. After normalizing the concentration, the RNA was reverse transcribed to cDNA and stored at −20 °C. Primers (Table S1) were designed with Primer 5.0 software, and sequence specificity was verified through BLAST analysis. The expressions of target genes were detected by qPCR, using GAPDH as the housekeeping gene. The reaction mixture had a volume of 20 μL, and the reaction conditions were as follows: denaturation at 94 °C for 30 s, annealing at 55 °C for 30 s (for 40 cycles), and extension at 72 °C for 5 min. The data were treated with a $2^{-\Delta\Delta Ct}$ relative expression method.

2.6. ELISA Assay

The expression levels of lysozyme (LZ), alkaline phosphatase (ALP), complement component 3 (C3), complement component 4 (C4), immunoglobulin T (IgT), immunoglobulin D (IgD), and immunoglobulin M (IgM) were detected by ELISA kits (Mmbio, China). ELISA kits purchased from Andy Gene (Beijing, China) were used to detect the levels of interleukin-6 (IL-6), interleukin-1β (IL-1β), and tumor necrosis factor-α (TNF-α). The intestinal tissue was homogenized in precooled PBS at a ratio of 1/9 (*w*/*v*) and centrifuged at 7500 rpm for 15 min. Then, the supernatant was collected, and the remaining operations were carried out by the instructions of ELISA kits. A microplate reader was used to detect the absorbance at a wavelength of 450 nm.

2.7. Western Blotting Assay

The intestinal tissue and cells were homogenized in lysis buffer at a ratio of 1/9 (*w*/*v*). The supernatant was collected by centrifugation after remaining on ice for 30 min. The protein concentration was determined by the BCA kit, adjusting to 5 μg/μL. Protein samples were separated by appropriate concentrations of SDS-PAGE (Servicebio, Wuhan, China) at 80 V for 30 min and 120 V for 60 min and transferred to NC membranes at 200 mA. The NC membranes were blocked in 5% skimmed milk for 2 h at room temperature and incubated with the primary antibody (Table S2) at 4 °C overnight. The corresponding secondary antibodies were used to incubate for 2 h at room temperature. Finally, the BCL luminescent solution was used to observe the membranes, and gray analysis was performed.

2.8. Protein–Protein Interaction (PPI) Analysis and Radar Plot

To explore the link between intestinal immune molecules and tight junctions, the STRING database was used to establish a PPI network. This database provides possible potential interactions between the encoded proteins. Enter the name of related genes or proteins, and the website will display the interaction of the related proteins [21].

Radar plot is a simple multivariate graphic method. It allows the visual integration of measured biomarkers. The data of multiple indexes are mapped to coordinate axes, which start from the same center point and end at the periphery of the circle. Connecting the points of the same treatment group forms a radar plot [22]. Hiplot tools were used to create a radar plot in this study [23].

2.9. Statistical Analysis

All data in this experiment were analyzed with GraphPad Prism 8.0.1 and SPSS 26.0 software. The experiments were repeated at least three times. Shapiro–Wilk was used to test the normal distribution of the data, and Brown–Forsythe was used to perform the homogeneity of variance. The data were performed as the mean ± standard deviation (mean ± SD). A one-way analysis of variance (ANOVA) was performed to analyze the differences between groups. Furthermore, $p < 0.05$ indicates significant differences.

3. Results

3.1. TP Relieves LPS-Induced Intestinal Tissue Damage

Common carp were given different treatments after seven days of adaptive feeding. Four groups were set up to observe whether TP has positive effects on the intestines under LPS challenge (Figure 1A). Histopathological staining was used to assess tissue damage, and the results are shown in Figure 1B. The intestinal mucosa was intact, and the villi were arranged in order in the CG and TP groups. Severe intestinal injury occurred in the LPS group, including loss and abscission of intestinal villi. Remarkably, the TL group showed a complete histological structure, and the intestinal villi were intact. Moreover, intestinal morphology indexes were also detected. The mucosal fold height and muscular layer thickness of the LPS group were significantly lower, which were increased in the TL group (Figure S2).

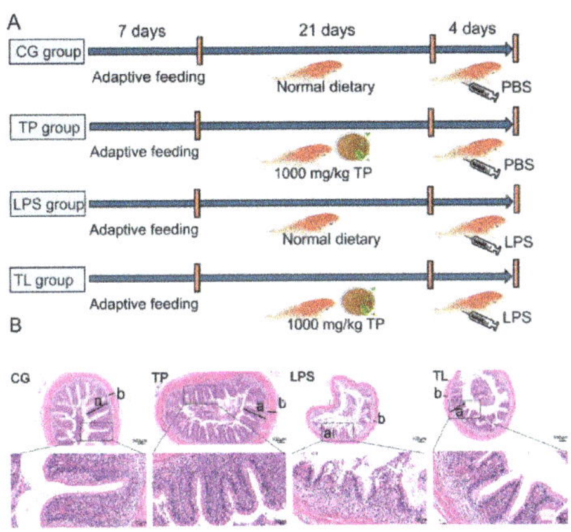

Figure 1. Effects of dietary TP on the intestinal pathological damage challenge by LPS in common carp. (**A**) Schematic diagram of the grouping and treatment on common carp. (**B**) Histopathological staining on the intestines of common carp. a represented the height of the mucosal fold, and b represented the thickness of the muscular layer. The scale bars were 100 µm and 20 µm.

3.2. TP Enhances the Intestinal Immune Ability

In response to stimulation, the intestinal immune system could be activated via increasing amounts of immune index, such as LZ, ALP, C3, C4, and Igs [24]. Figure 2 shows the changes in immune index after TP and LPS treatment. The mRNA levels of *mucin-2* (*muc-2*), *ALP*, *LZ*, *C3*, *C4*, *IgT*, *IgD*, and *IgM* were increased in the TL group compared with that in the LPS group (Figure 2A). The enzyme activity of ALP and LZ was significantly increased after LPS stimulation than in the CG. However, the TP group showed higher enzyme activity than the LPS group (Figure 2B,C). LPS stimulated the levels of C3 and C4, with further increasing levels of C3 and C4 in the TL group (Figure 2D,E). Moreover, the levels of IgT, IgM, and IgD have different degrees of increase after LPS treatment compared to the CG; the TL group showed higher levels of Igs compared with the LPS group (Figure 2F–H). These results indicated that TP promoted the immune ability in the intestines of common carp.

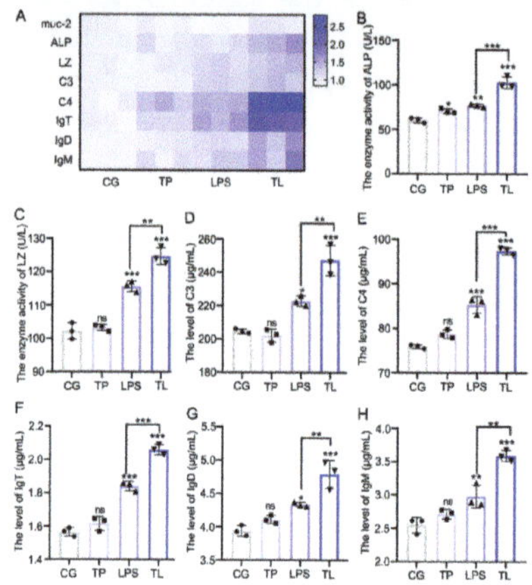

Figure 2. Effects of TP on the intestinal immunity parameters in common carp. (**A**) The mRNA levels of *muc-2*, *ALP*, *LZ*, *C3*, *C4*, *IgT*, *IgD*, and *IgM* in intestinal tissue. (**B**,**C**) The enzymatic activity of ALP and LZ in intestinal tissue. (**D**–**H**) The protein expression levels of C3, C4, IgT, IgD, and IgM in intestinal tissue. * $p \leq 0.05$, ** $p \leq 0.01$, *** $p \leq 0.001$, ns $p > 0.05$.

3.3. TP Relieves LPS-Induced Intestinal Inflammatory Injury

The inflammatory cytokines of the intestines and cells were detected by qPCR and ELISA. Compared with the CG, LPS significantly increased the mRNA expression levels of *IL-6*, *IL-1β*, and *TNF-α*. On the contrary, the TL group decreased the mRNA expression levels of these pro-inflammatory factors compared with the LPS group in intestinal tissue and cells (Figure 3A–F). The protein expression level of pro-inflammatory factors was consistent with the mRNA expression levels (Figure 3G–L). These results showed that TP alleviated the intestinal inflammatory injury caused by LPS in common carp.

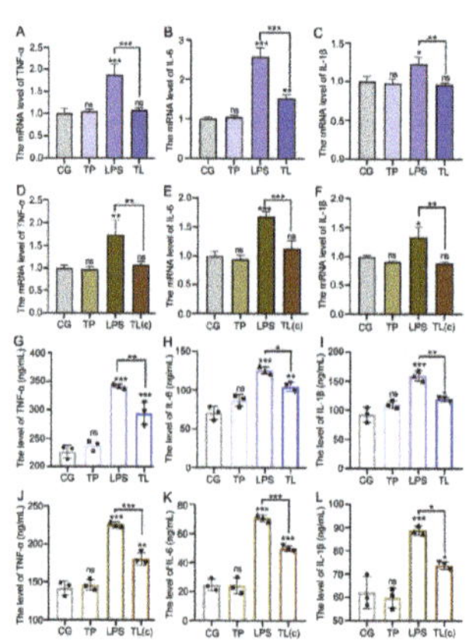

Figure 3. Effects of TP on LPS-induced intestinal inflammatory injury in common carp. (**A**–**C**) The mRNA levels of *IL-6*, *IL-1β*, and *TNF-α* in intestinal tissue. (**D**–**F**) The mRNA levels of *IL-6*, *IL-1β*, and *TNF-α* in intestinal epithelial cells. (**G**–**I**) The protein expression levels of IL-6, IL-1β, and TNF-α in intestinal tissue. (**J**–**L**) The protein expression levels of IL-6, IL-1β, and TNF-α in intestinal epithelial cells. * $p \leq 0.05$, ** $p \leq 0.01$, *** $p \leq 0.001$, ns $p > 0.05$.

3.4. TP Inhibits LPS-Induced Activation of the TLR2/NF-κB Pathway

To detect whether TP interacted with TLR2, the AutoDock tool was used to build the model of the TP-TLR2 complex through homology modeling and ligand docking. As shown in Figure 4A, TP mainly interacts with TLR2 through hydrogen bonds. TP formed hydrogen bonds with GLY-334, with hydrogen bond lengths of 1.9 Å and 2.0 Å. The binding energy of TP and TLR2 was −7.9 kcal/mol, which proved that TP had a strong binding activity with TLR2. Subsequently, the expression levels of inflammatory pathway-related factors were detected. The mRNA levels of *TLR2*, *MyD88*, *NF-κB p65*, and *IκBα* increase remarkably after being treated with LPS compared with the CG in tissue and cells. However, these indicators significantly decreased in the TL group (Figure 4B,C). The protein expression levels of TLR2, MyD88, *p*-p65, and *p*-IκBα showed a similar trend with mRNA levels (Figures 4D,E and S3). These results indicated that TP significantly inhibited the activation of the TLR2/NF-κB pathway.

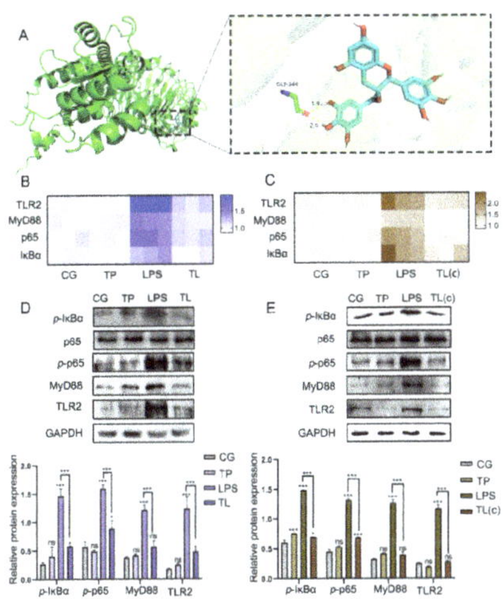

Figure 4. Effects of TP on the TLR2/NF-κB pathway in the intestines of common carp. (**A**) Molecular docking was used to detect the interaction between TP and TLR2. (**B**) The mRNA levels of *TLR2*, *MyD88*, *NF-κB p65*, and *IκBα* in intestinal tissue. (**C**) The mRNA levels of *TLR2*, *MyD88*, *NF-κB p65*, and *IκBα* in intestinal epithelial cells. (**D**,**E**) The protein expression levels of *p*-IκBα, p65, *p*-p65, MyD88, and TLR2 in intestinal tissue and cells. * $p \leq 0.05$, *** $p \leq 0.001$, ns $p > 0.05$.

3.5. TP Relieves LPS-Caused Destruction of Tight Junction

The PPI network showed a close association between intestinal immune factors and tight junctions (Figure 5A). Subsequently, the mRNA levels of tight junction-related factors were examined. The mRNA levels of *claudin-1*, *claudin-2*, *claudin-5*, *occludin*, *ZO-1*, and *ZO-3* were decreased in the LPS group compared with the CG. The TL group increased the mRNA levels of these genes compared with the LPS group in tissue and cells (Figure 5B,C). The expression levels of occludin and claudin-1 showed similar trends as the mRNA levels (Figures 5D,E and S3). The MLCK-MLC pathway played a pivotal role in the tight junction of the intestines. In this study, LPS significantly increased the mRNA levels of *MLCK* and *MLC*, which were significantly decreased in the TL group (Figure 5B,C). The expression levels of *p*-MLC were consistent with the trend in the mRNA levels (Figure 5D,E). These results indicated that TP protected the tight junction in the intestines of common carp.

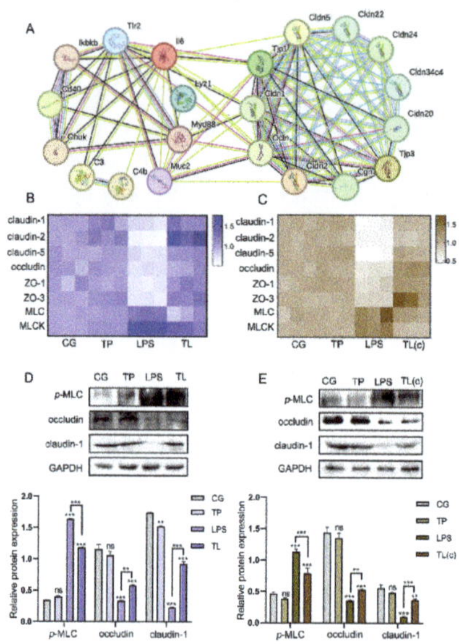

Figure 5. Effects of TP on the intestinal tight junction parameters in common carp. (**A**) The PPI network between intestinal immune factors and tight junctions. (**B**) The mRNA levels of *claudin-1*, *claudin-2*, *claudin-5*, *occludin*, *ZO-1*, *ZO-3*, *MLC*, and *MLCK* in intestinal tissue. (**C**) The mRNA levels of *claudin-1*, *claudin-2*, *claudin-5*, *occludin*, *ZO-1*, *ZO-3*, *MLC*, and *MLCK* in intestinal epithelial cells. (**D**,**E**) The protein expression levels of *p*-MLC, occludin, and claudin-1. ** $p \leq 0.01$, *** $p \leq 0.001$, ns $p > 0.05$.

3.6. Radar Plot Analysis of TP on the Mechanism of LPS-Induced Intestinal Injury

A radar plot was used to visually integrate biomarkers that were detected in this study. Biomarkers accounted for a large proportion of inflammation and tight junction disruption-related pathways in the LPS group. However, biomarkers were mainly concentrated in immune factors and tight junction proteins in the TL group (Figure 6). These results showed that TP activated intestinal immunity and protected intestinal tight junctions to alleviate intestinal injury.

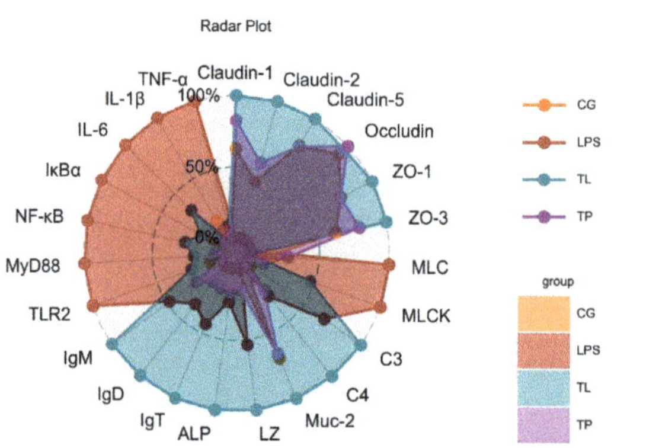

Figure 6. Radar plot of TP on the multiple biomarkers of intestinal damage in common carp. The data of the intestinal immune barrier, inflammation, and tight junction are mapped to coordinate axes, which start from the same center point and end at the periphery of the circle. Orange represents the CG; purple represents the TP group; red represents the LPS group; and blue represents the TL group.

4. Discussion

Tea polyphenol, absorbed through the intestine, has great potential in aquaculture. TP can regulate defense mechanisms to indirectly maintain host health [25]. Studies have demonstrated the immunomodulatory effects of tea on fish, such as *Oncorhynchus mykiss*, *Labeo rohita*, and *Epinephelus bruneus* [26–28]. LPS plays an important role in pathological damage, inflammation, and immune activation in fish [29]. Therefore, this study established an LPS-induced intestinal injury model to explore the protective effect of TP on the intestines of common carp. H&E results showed that LPS caused severe histopathological damage in the intestine, manifested as shortening and fragmentation of intestinal villi and reduction in muscular layer thickness. TP significantly relieved the intestinal injury, showing intact intestinal villi. Mucosal fold height and muscular layer thickness were significantly higher than those of the LPS group. These results indicated that TP could alleviate LPS-induced intestinal tissue injury.

The immune function of the intestines occupies a key position in the immune defense of teleost. The immune system can directly or indirectly affect muc-2 expression [30]. In this study, LPS induced a decrease in intestinal muc-2 gene expression, while TP could promote the expression of this gene. TP can increase LZ and ALP levels and play an immunostimulatory role in hybrid groupers [15]. Studies have pointed out that dietary Camellia sinensis increases the mRNA level of *LZ* to enhance the fish's immune system [26]. In this study, TP treatment significantly increased the expression levels of *LZ* and *ALP*, which is consistent with the above studies to a certain extent. Complement is an important component in fighting microbial organisms. C3, the central mediator of the complement system, plays an immune surveillance role [31]. This study found that the addition of TP significantly increased the expression levels of C3 and C4, suggesting that TP stimulates the body to resist LPS challenge. Previous studies have shown that dietary *Jasonia glutinosa* increases the levels of IgM and complement activity in body fluids to stimulate the immune activity of gilthead seabream [32]. In this study, the TL group significantly increased the expression level of IgM, IgD, and IgT, suggesting that TP activated intestinal mucosal immunity. This study reported that TP can improve the immune capacity by regulating the innate and adaptive immune responses in the intestines of common carp.

Inflammatory response and immune regulation are two closely related processes that work together to maintain the stability of the body's internal environment. Cytokines are playing an important role in regulating intestinal barrier function and intercellular interaction in the intestinal immune system [33]. This study found that TP inhibited the expression levels of inflammatory factors TNF-α, IL-6, and IL-1β, which were increased by LPS. TLRs are major components in the regulation of innate and adaptive immunity [34]. Studies have shown that dietary citrus flavonoids improve immune function by regulating TLR2/NF-κB signaling [35]. Since we found that TP could increase the level of intestinal immunity and inhibit the secretion of inflammatory factors, we hypothesized that TP might exert an immunomodulatory effect by binding to TLR2. Molecular docking showed that the binding energy of TP and TLR2 was -7.9 kcal/mol, indicating that they had strong binding activity. Moreover, the mRNA and protein levels of TLR2 were inhibited by TP in this study. Some studies have found that LPS stimulates NF-κB activation through the TLR2/MyD88 pathway [36]. In this study, LPS also activated the TLR2/MyD88/NF-κB pathway to cause an intestinal inflammatory response. However, TP can significantly inhibit the expression of related proteins, thereby exerting a protective effect. Previous studies have shown that epigallocatechin-3-gallate, the active component of TP, downregulates NF-κB and TLR signaling to inhibit the production of inflammatory cytokines [37]. The results suggested that TP alleviates LPS-induced intestinal inflammation injury by inhibiting the TLR2/NF-κB pathway. Additionally, the PPI network showed that intestinal immune-related factors were closely associated with tight junctions. NF-κB affects the tight junction barrier by regulating MLCK/*p*-MLC [38]. Studies have shown that *Scytosiphon lomentaria* fucoidan inhibits the TLR4/NF-κB/MLCK pathway to improve tight junction protein levels, protecting tight junction integrity in dietary fiber-deficient mice [39]. In this study, TP inhibited LPS-induced the expression of MLCK and *p*-MLC. Studies have pointed out that luteolin alleviates ethanol-induced tight junction destruction by inhibiting NF-κB nuclear translocation, activating MLCK, and promoting MLC phosphorylation [40]. TP can inhibit TLR4/MLCK signal transduction and protect the pulmonary tracheal epithelial tight junction [41]. MLCK is a key component in regulating tight junction proteins. The upregulation of MLCK can inhibit the mRNA levels of claudins, ZO-1, and occludin [42]. This study revealed that TP alleviated LPS caused by the downregulation of claudins, occludin, and ZO-1 in intestinal tissue and cells. Similar to the present study, dietary TP could inhibit the expression of MLCK and promote the expression of tight junction factors to protect the intestines in the yellow drum [43]. This study found that TP significantly inhibited LPS-induced activation of the NF-κB/MLCK signal to alleviate the disruption of tight junctions in intestinal tissue and cells.

5. Conclusions

TP alleviated LPS-induced intestinal injury, protecting the intestinal integrity of common carp. TP could bind to TLR2 to inhibit the expression of NF-κB, activating the intestinal immune capacity. TP further prevented LPS-induced disruption of tight junctions by inhibiting NF-κB/MLCK signal activation. Our study could be valuable for understanding the protective effects of TP on the fish intestines, which supports the regular addition of TP to the diet of fish.

Supplementary Materials: The following supporting information can be downloaded at: https://www.mdpi.com/article/10.3390/ani15030387/s1, Figure S1: Effects of tea polyphenol (TP) on the viability of primary intestinal epithelial cells; Figure S2: Intestinal morphology indexes of common carp with different treatment; Figure S3: Western blotting images; Table S1: RT-PCR primer information; Table S2: Antibodies used for Western blotting.

Author Contributions: Investigation, J.W., Z.L. and J.L.; methodology, J.Y. and Y.X.; supervision, X.G.; writing—original draft, M.Q.; writing—review and editing, B.Z. and X.G. All authors have read and agreed to the published version of the manuscript.

Funding: This research was supported by the Natural Science Foundation of Heilongjiang (YQ2021C022) and the Postdoctoral Scientific Research Developmental Fund of Heilongjiang (LBH-Q19080).

Institutional Review Board Statement: This study was conducted in accordance with the Institutional Animal Care and Use Committee of Northeast Agriculture University (NEAU2023046).

Informed Consent Statement: Not applicable.

Data Availability Statement: The original contributions presented in the study are included in the article/Supplementary Materials, further inquiries can be directed to the corresponding authors.

Conflicts of Interest: The authors declare no conflicts of interest.

References

1. Shang, X.; Yu, P.; Yin, Y.; Zhang, Y.; Lu, Y.; Mao, Q.; Li, Y. Effect of selenium-rich Bacillus subtilis against mercury-induced intestinal damage repair and oxidative stress in common carp. *Comp. Biochem. Physiol. C Toxicol. Pharmacol.* **2021**, *239*, 108851. [CrossRef] [PubMed]
2. Chen, W.; Gao, S.; Huang, Y.; Chang, K.; Zhao, X. Addition of Chlorella sorokiniana meal in the diet of juvenile rainbow trout (*Oncorhynchus mykiss*): Influence on fish growth, gut histology, oxidative stress, immune response, and disease resistance against Aeromonas salmonicida. *Fish Shellfish Immunol.* **2022**, *129*, 243–250. [CrossRef] [PubMed]
3. Zhang, C.N.; Li, X.F.; Jiang, G.Z.; Zhang, D.D.; Tian, H.Y.; Li, J.Y.; Liu, W.B. Effects of dietary fructooligosaccharide levels and feeding modes on growth, immune responses, antioxidant capability and disease resistance of blunt snout bream (*Megalobrama amblycephala*). *Fish Shellfish Immunol.* **2014**, *41*, 560–569. [CrossRef] [PubMed]
4. Kong, W.G.; Yu, Y.Y.; Dong, S.; Huang, Z.Y.; Ding, L.G.; Cao, J.F.; Dong, F.; Zhang, X.T.; Liu, X.; Xu, H.Y.; et al. Pharyngeal Immunity in Early Vertebrates Provides Functional and Evolutionary Insight into Mucosal Homeostasis. *J. Immunol.* **2019**, *203*, 3054–3067. [CrossRef] [PubMed]
5. Nguyen, T.M.; Agbohessou, P.S.; Nguyen, T.H.; Tran Thi, N.T.; Kestemont, P. Immune responses and acute inflammation in common carp *Cyprinus carpio* injected by *E.coli* lipopolysaccharide (LPS) as affected by dietary oils. *Fish Shellfish Immunol.* **2022**, *122*, 1–12. [CrossRef] [PubMed]
6. Gao, X.J.; Tang, B.; Liang, H.H.; Yi, L.; Wei, Z.G. Selenium deficiency induced an inflammatory response by the HSP60—TLR2-MAPKs signalling pathway in the liver of carp. *Fish Shellfish Immunol.* **2019**, *87*, 688–694. [CrossRef] [PubMed]
7. O'Neill, L.A.; Golenbock, D.; Bowie, A.G. The history of Toll-like receptors—Redefining innate immunity. *Nat. Rev. Immunol.* **2013**, *13*, 453–460. [CrossRef]
8. Barnabei, L.; Laplantine, E.; Mbongo, W.; Rieux-Laucat, F.; Weil, R. NF-κB: At the Borders of Autoimmunity and Inflammation. *Front. Immunol.* **2021**, *12*, 716469. [CrossRef] [PubMed]
9. Jiang, J.; Lu, X.; Dong, L.X.; Peng, D.; Zhang, J.M.; Tian, J.; Wen, H.; Jiang, M. Dietary cholesterol intervention could alleviate the intestinal injury of *Oreochromis niloticus* induced by plant-based diet via the intestinal barriers. *Fish Shellfish Immunol.* **2024**, *150*, 109621. [CrossRef] [PubMed]
10. Liu, S.; Li, J.; Kang, W.; Li, Y.; Ge, L.; Liu, D.; Liu, Y.; Huang, K. Aflatoxin B1 Induces Intestinal Barrier Dysfunction by Regulating the FXR-Mediated MLCK Signaling Pathway in Mice and in IPEC-J2 Cells. *J. Agric. Food Chem.* **2023**, *71*, 867–876. [CrossRef] [PubMed]
11. Chen, S.; Zhu, J.; Chen, G.; Zuo, S.; Zhang, J.; Chen, Z.; Wang, X.; Li, J.; Liu, Y.; Wang, P. 1,25-Dihydroxyvitamin D3 preserves intestinal epithelial barrier function from TNF-α induced injury via suppression of NF-kB p65 mediated MLCK-P-MLC signaling pathway. *Biochem. Biophys. Res. Commun.* **2015**, *460*, 873–878. [CrossRef]
12. Zhao, X.; Shi, X.; Liu, Q.; Li, X. Tea polyphenols alleviates acetochlor-induced apoptosis and necroptosis via ROS/MAPK/NF-κB signaling in *Ctenopharyngodon idellus* kidney cells. *Aquat. Toxicol.* **2022**, *246*, 106153. [CrossRef] [PubMed]
13. Yan, Z.; Zhong, Y.; Duan, Y.; Chen, Q.; Li, F. Antioxidant mechanism of tea polyphenols and its impact on health benefits. *Anim. Nutr.* **2020**, *6*, 115–123. [CrossRef] [PubMed]
14. Cheng, Z.; Wang, Y.; Li, B. Dietary Polyphenols Alleviate Autoimmune Liver Disease by Mediating the Intestinal Microenvironment: Challenges and Hopes. *J. Agric. Food Chem.* **2022**, *70*, 10708–10737. [CrossRef]
15. Pan, S.; Yan, X.; Li, T.; Suo, X.; Liu, H.; Tan, B.; Huang, W.; Yang, Y.; Zhang, H.; Dong, X. Impacts of tea polyphenols on growth, antioxidant capacity and immunity in juvenile hybrid grouper (*Epinephelus fuscoguttatus* ♀× *E. lanceolatus* ♂) fed high-lipid diets. *Fish Shellfish Immunol.* **2022**, *128*, 348–359. [CrossRef] [PubMed]
16. Xu, R.; Han, F.X.; Wang, H.R.; Wang, J.J.; Cai, Z.L.; Guo, M.Y. Tea polyphenols alleviate TBBPA-induced inflammation, ferroptosis and apoptosis via TLR4/NF-κB pathway in carp gills. *Fish Shellfish Immunol.* **2024**, *146*, 109382. [CrossRef] [PubMed]
17. Zhu, H.; Gao, M.; Sun, W.; Liu, H.; Xu, S.; Li, X. ROS/ER stress contributes to trimethyltin chloride-mediated hepatotoxicity; Tea polyphenols alleviate apoptosis and immunosuppression. *Comp. Biochem. Physiol. C Toxicol. Pharmacol.* **2023**, *263*, 109505. [CrossRef]
18. Yu, Z.; Zhao, Y.Y.; Jiang, N.; Zhang, A.Z.; Li, M.Y. Bioflocs attenuates lipopolysaccharide-induced inflammation, immunosuppression and oxidative stress in *Channa argus*. *Fish Shellfish Immunol.* **2021**, *114*, 218–228. [CrossRef]
19. Wang, Y.; Yan, X.; Fu, L. Effect of selenium nanoparticles with different sizes in primary cultured intestinal epithelial cells of crucian carp, *Carassius auratus* gibelio. *Int. J. Nanomed.* **2013**, *8*, 4007–4013. [CrossRef]
20. Jiang, J.; Shi, D.; Zhou, X.Q.; Hu, Y.; Feng, L.; Liu, Y.; Jiang, W.D.; Zhao, Y. In vitro and in vivo protective effect of arginine against lipopolysaccharide induced inflammatory response in the intestine of juvenile Jian carp (*Cyprinus carpio* var. Jian). *Fish Shellfish Immunol.* **2015**, *42*, 457–464. [CrossRef] [PubMed]

21. Gan, J.; Ma, H.; Ma, Y.; Zhou, M.; Li, Y.; Yan, W.; Dong, Z. Silybin prevented avermectin-induced cardiotoxicity in carp by modulating oxidative stress, inflammation, endoplasmic reticulum stress, mitochondrial apoptosis, and autophagy. *Fish Shellfish Immunol.* **2024**, *150*, 109624. [CrossRef] [PubMed]
22. Beliaeff, B.; Burgeot, T. Integrated biomarker response: A useful tool for ecological risk assessment. *Environ. Toxicol. Chem.* **2002**, *21*, 1316–1322. [CrossRef] [PubMed]
23. Li, J.; Miao, B.; Wang, S.; Dong, W.; Xu, H.; Si, C.; Wang, W.; Duan, S.; Lou, J.; Bao, Z.; et al. Hiplot: A comprehensive and easy-to-use web service for boosting publication-ready biomedical data visualization. *Brief. Bioinform.* **2022**, *23*, bbac261. [CrossRef]
24. Wu, J.; Yang, W.; Song, R.; Li, Z.; Jia, X.; Zhang, H.; Zhang, P.; Xue, X.; Li, S.; Xie, Y.; et al. Dietary Soybean Lecithin Improves Growth, Immunity, Antioxidant Capability and Intestinal Barrier Functions in Largemouth Bass Micropterus salmoides Juveniles. *Metabolites* **2023**, *13*, 512. [CrossRef] [PubMed]
25. Mithul Aravind, S.; Wichienchot, S.; Tsao, R.; Ramakrishnan, S.; Chakkaravarthi, S. Role of dietary polyphenols on gut microbiota, their metabolites and health benefits. *Food Res. Int.* **2021**, *142*, 110189. [CrossRef] [PubMed]
26. Debbarma, S.; Acharya, A.; Mangang, Y.A.; Monsang, S.J.; Choudhury, T.G.; Parhi, J.; Pandey, P.K. Immune-biochemical response and immune gene expression profiling of Labeo rohita fingerlings fed with ethanolic tea leaf extracts and its survivability against *Aeromonas hydrophila* infection. *Fish Shellfish Immunol.* **2022**, *130*, 520–529. [CrossRef]
27. Harikrishnan, R.; Balasundaram, C.; Heo, M.S. Influence of diet enriched with green tea on innate humoral and cellular immune response of kelp grouper (*Epinephelus bruneus*) to *Vibrio carchariae* infection. *Fish Shellfish Immunol.* **2011**, *30*, 972–979. [CrossRef]
28. Sheikhzadeh, N.; Nofouzi, K.; Delazar, A.; Oushani, A.K. Immunomodulatory effects of decaffeinated green tea (*Camellia sinensis*) on the immune system of rainbow trout (*Oncorhynchus mykiss*). *Fish Shellfish Immunol.* **2011**, *31*, 1268–1269. [CrossRef] [PubMed]
29. Swain, P.; Nayak, S.K.; Nanda, P.K.; Dash, S. Biological effects of bacterial lipopolysaccharide (endotoxin) in fish: A review. *Fish Shellfish Immunol.* **2008**, *25*, 191–201. [CrossRef] [PubMed]
30. Inaba, R.; Vujakovic, S.; Bergstrom, K. The gut mucus network: A dynamic liaison between microbes and the immune system. *Semin. Immunol.* **2023**, *69*, 101807. [CrossRef] [PubMed]
31. Wang, L.; Wu, D.; Fan, Z.; Li, H.; Li, J.; Zhang, Y.; Xu, Q.; Wang, G.; Zhu, Z. Effect of Yucca schidigera extract on the growth performance, intestinal antioxidant status, immune response, and tight junctions of mirror carp (*Cyprinus carpio*). *Fish Shellfish Immunol.* **2020**, *103*, 211–219. [CrossRef] [PubMed]
32. Espinosa, C.; García Beltrán, J.M.; Messina, C.M.; Esteban, M. Effect of Jasonia glutinosa on immune and oxidative status of gilthead seabream (*Sparus aurata* L.). *Fish Shellfish Immunol.* **2020**, *100*, 58–69. [CrossRef]
33. Wang, J.; Li, M.; Gao, Y.; Li, H.; Fang, L.; Liu, C.; Liu, X.; Min, W. Effects of Exopolysaccharides from Lactiplantibacillus plantarum JLAU103 on Intestinal Immune Response, Oxidative Stress, and Microbial Communities in Cyclophosphamide-Induced Immunosuppressed Mice. *J. Agric. Food Chem.* **2022**, *70*, 2197–2210. [CrossRef] [PubMed]
34. de Lavergne, M.; Maisonneuve, L.; Podsypanina, K.; Manoury, B. The role of the antigen processing machinery in the regulation and trafficking of intracellular -Toll-like receptor molecules. *Curr. Opin. Immunol.* **2023**, *84*, 102375. [CrossRef]
35. Chen, C.; Feng, F.; Qi, M.; Chen, Q.; Tang, W.; Diao, H.; Hu, Z.; Qiu, Y.; Li, Z.; Chu, Y.; et al. Dietary Citrus Flavonoids Improved Growth Performance and Intestinal Microbiota of Weaned Piglets via Immune Function Mediated by TLR2/NF-κB Signaling Pathway. *J. Agric. Food Chem.* **2024**, *72*, 16761–16776. [CrossRef] [PubMed]
36. Mohammadi, G.; Karimi, A.A.; Hafezieh, M.; Dawood, M.A.O.; Abo-Al-Ela, H.G. Pistachio hull polysaccharide protects Nile tilapia against LPS-induced excessive inflammatory responses and oxidative stress, possibly via TLR2 and Nrf2 signaling pathways. *Fish Shellfish Immunol.* **2022**, *121*, 276–284. [CrossRef]
37. Shen, H.; Wu, N.; Liu, Z.; Zhao, H.; Zhao, M. Epigallocatechin-3-gallate alleviates paraquat-induced acute lung injury and inhibits upregulation of toll-like receptors. *Life Sci.* **2017**, *170*, 25–32. [CrossRef] [PubMed]
38. Zhong, J.; Yu, R.; Zhou, Q.; Liu, P.; Liu, Z.; Bian, Y. Naringenin prevents TNF-α-induced gut-vascular barrier disruption associated with inhibiting the NF-κB-mediated MLCK/p-MLC and NLRP3 pathways. *Food Funct.* **2021**, *12*, 2715–2725. [CrossRef] [PubMed]
39. Jia, J.; Zheng, W.; Tang, S.; Song, S.; Ai, C. Scytosiphon lomentaria fucoidan ameliorates DSS-induced colitis in dietary fiber-deficient mice via modulating the gut microbiota and inhibiting the TLR4/NF-κB/MLCK pathway. *Int. J. Biol. Macromol.* **2023**, *253*, 127337. [CrossRef]
40. Yuan, J.; Che, S.; Zhang, L.; Ruan, Z. Reparative Effects of Ethanol-Induced Intestinal Barrier Injury by Flavonoid Luteolin via MAPK/NF-κB/MLCK and Nrf2 Signaling Pathways. *J. Agric. Food Chem.* **2021**, *69*, 4101–4110. [CrossRef]
41. Li, X.; Liu, Z.; Gao, T.; Liu, W.; Yang, K.; Guo, R.; Li, C.; Tian, Y.; Wang, N.; Zhou, D.; et al. Tea Polyphenols Protects Tracheal Epithelial Tight Junctions in Lung during Actinobacillus pleuropneumoniae Infection via Suppressing TLR-4/MAPK/PKC-MLCK Signaling. *Int. J. Mol. Sci.* **2023**, *24*, 11842. [CrossRef] [PubMed]

42. Feng, L.; Li, S.Q.; Jiang, W.D.; Liu, Y.; Jiang, J.; Wu, P.; Zhao, J.; Kuang, S.Y.; Tang, L.; Tang, W.N.; et al. Deficiency of dietary niacin impaired intestinal mucosal immune function via regulating intestinal NF-κB, Nrf2 and MLCK signaling pathways in young grass carp (*Ctenopharyngodon idella*). *Fish Shellfish Immunol.* **2016**, *49*, 177–193. [CrossRef]
43. Ma, S.; Wang, L.; Zeng, Y.; Tan, P.; Chen, R.; Hu, W.; Xu, H.; Xu, D. Reparative effect of different dietary additives on soybean meal-induced intestinal injury in yellow drum (*Nibea albiflora*). *Front. Immunol.* **2023**, *14*, 1296848. [CrossRef] [PubMed]

animals

MDPI

Article

Glycyrrhizin Alleviates the Damage Caused by Zearalenone and Protects the Glandular Stomach of Chickens

Tong Sun, Fuhan Wang, Man Qian, Jingjing Wang and Mengyao Guo *

College of Veterinary Medicine, Northeast Agricultural University, Harbin 150030, China; suntong@neau.edu.cn (T.S.); wangfuhan1997@sina.com (F.W.); qm18437916979@163.com (M.Q.); s230602012@neau.edu.cn (J.W.)
* Correspondence: authors: gmy@neau.edu.cn

Simple Summary: Zearalenone (ZEA) is a mycotoxin that widely contaminates feed. Glycyrrhizin acid (GA) is a traditional Chinese medicine extract of licorice. This study found that GA could alleviate the damage to chicken glands and stomachs caused by ZEA in vivo, reduce the levels of oxidative stress in the body, and reduce the number of instances of apoptosis and programmed necrosis. GA may inhibit apoptosis and programmed necrosis mediated by inflammation through the NFκB pathway and protect the glandular gastric tissue. These results provide indirection for toxicological studies of ZEA and therapeutic effects of GA and help further explore the toxic effects of mycotoxins on the digestive tract.

Abstract: Zearalenone (ZEA) is a kind of mycotoxin that widely contaminates food and feed and poses a threat to poultry farming. As a natural extract, glycyrrhizin acid (GA) has antioxidant, antibacterial, and anti-inflammatory effects. Although studies have revealed the toxic effects of ZEA on the liver, the mechanism by which GA reduces ZEA's toxic protective glandular stomach remains unclear. In order to study the therapeutic effect of GA on tissue damage caused by ZEA, we conducted in vivo and in vitro experiments to compare the expression of inflammation, oxidative stress, apoptosis, and necrosis. The results showed that ZEA can induce inflammation in tissues and cells, inducing apoptosis and necrosis. In addition, GA can alleviate the toxic effects caused by ZEA and protect cells. Dietary GA significantly increased the antioxidant capacity of glandulae and inhibited the overexpression of NFκB/IκB-α and its mediated inflammatory response. Moreover, GA decreased the expression of pro-apoptotic factors and necrosis factors, thereby alleviating apoptosis and necrosis of chicken glandular stomach cells. At present, the mechanism of ZEA damage to livers and lungs has been confirmed by studies. However, there have been no studies on GA alleviating the damage caused by ZEA to the glandular stomach. Therefore, the purpose of this study was to explore the mechanism of GA alleviating the damage caused by ZEA in the glandular stomach through in vivo and in vitro experimental comparison. The results may provide some reference for the solution of feed contamination.

Keywords: zearalenone; glycyrrhizic acid; inflammation; oxidative stress; apoptosis; necrosis

1. Introduction

Zearalenone, a mycotoxin [1,2], widely pollutes various grains around the world, such as wheat, barley, corn, sorghum, rye, rice, and other grains and feeds [3]. ZEA causes varying degrees of harm to humans and animals [4] by changing the morphology and function of tissues and organs. The toxic effects of ZEA can also be aggravated in combination with other toxins [5]. ZEA impairs intestinal integrity and function, leading to pathological symptoms and tissue damage [6]. ZEA is widely absorbed by mice, pigs, poultry, and humans through the oral consumption of contaminated grains, foods, and feeds. Studies have shown that ZEA is rapidly absorbed through the gastrointestinal tract after oral ingestion. Feed contamination has been shown to cause intestinal inflammation and cecum injury in pigs [7,8], and a ZEA-contaminated diet was shown to reduce rumen PH and impair their function [9,10]. ZEA is widely distributed in various tissues and organs of animals and is slowly metabolized to clear the body. In addition, ZEA can be deposited in animals or animal products through feed and continue to cause harm [11].

The nutritional value of feed contaminated with mycotoxin is greatly reduced, which affects animal health, can cause tissue inflammation, and reduces antioxidant enzyme activity. ZEA, as an estrogen toxoid, affects the expression of inflammatory factors in poultry serum and ROS expression in vivo [12,13]. Studies have shown that ZEA induces apoptosis and inflammation of jejunal epithelial cells, leading to jejunal tissue damage [14]. Feeding low-dose ZEA-contaminated feed for a long time can cause intestinal inflammation in piglets. Diets contaminated with a high dose of ZEA can affect the antioxidant capacity of IPEC-J2 cells in the small intestine of pigs, resulting in ROS accumulation and mitochondrial apoptosis [15]. ZEA can induce hepatocyte apoptosis and affect liver function by activating the NFκB-signaling pathway by regulating inflammation-related factors [16–18]. Feeding chickens with different levels of ZEA contamination, we found that ZEA exposure significantly increased malondialdehyde (MDA) levels and inhibited catalase (CAT) activity, suggesting that ZEA induces tissue damage in chicks by regulating oxidative stress [19]. The results show that T-2 toxin can significantly affect the activity of digestive enzymes, causing intestinal mucosal inflammation, necrosis, and other symptoms. It enters the gastrointestinal tract through the mouth and begins to regulate genes involved in oxidative stress, inflammation, apoptosis, cancer, and other related genes, resulting in glandular and gastric damage [20].

At present, no specific medicine exists for damage caused to poultry by feed contamination by ZEA. Licorice is derived from the roots and rhizomes of the licorice plant and shows good antioxidant, antifungal, and anti-inflammatory effects, making it a widely developed Chinese medicine therapeutic agent. Glycyrrhizin, a compound extracted from licorice root, is one of licorice's most potent therapeutic ingredients [21,22]. GA mainly inhibits the expression of the nuclear factor kappa B (NFκB) pathway and then inhibits the encoding of important genes such as inflammatory cytokines, anti-apoptotic factors, and anti-necrosis factors so as to achieve anti-inflammatory effects. To date, no studies have been conducted on GA alleviating the damage caused by ZEA to the body. Therefore, this study mainly describes the mechanism of ZEA damage to the chicken glandular stomach and how GA alleviates panapoptosis caused by ZEA in the body, providing a certain reference for a solution to feed contamination.

2. Materials and Methods

2.1. *Animals and Groups*

Animal experiments were conducted in accordance with the guidelines of the Animal Care and Use Committee of Northeast Agricultural University. Fifty 7-day healthy chickens were randomly divided into five groups: the control group (Con group), the ZEA group, the ZEA+LGA group, the ZEA+MGA group, and the ZEA+HGA group. Each chicken is kept in a separate cage. The control group (Con) was fed common chicken feed and water. Chicken feed is bought commercially. It is composed of corn, soybean meal, bran, calcium bicarbonate, and sodium chloride. The feed does not contain fungal inhibitors. The feed purchased is verified by the kit to be ZEA-free. The ZEA group was fed 2 mg/kg of ZEA and ordinary chicken feed every 24 h. ZEA and GA were acquired from Shanghai Aladdin Biochemical Technology Co., LTD (Shanghai, China). The ZEA+LGA group was provided 2 mg/kg of ZEA every 24 h, and the chickens were fed diets containing 50 mg/kg of GA. The ZEA+MGA group was provided 2 mg/kg of ZEA every 24 h and fed chicken feed with 100 mg/kg of GA content. The ZEA+HGA group was provided 2 mg/kg of ZEA every 24 h, and the chickens were fed diets containing 150 mg/kg of GA. ZEA is mixed with feed. The stability of ZEA intake is ensured by maintaining the daily feed intake of each chicken. All the chickens were allowed free access to food and water and were kept at room temperature. After 21 days of exposure to ZEA, all animals were anesthetized and killed. We collected glandular gastric tissue. Part of the fresh glandular gastric tissue was fixed using paraformaldehyde. Other glandular gastric tissues were quickly cryogenically preserved in liquid nitrogen for subsequent analysis.

2.2. Histopathology Staining

The glandular stomach tissue was fixed with 4% paraformaldehyde, and 4 μm sections were prepared by embedding the tissue with paraffin wax. After dewaxing the sections, they were stained with hematoxylin for 5 min, rinsed with eosin for 3 min, and sealed with neutral glue after dehydration. The slices were then observed under a microscope (digital tissue section scanner, 3DHISTECH, Budapest, Hungary), and stained images were obtained using an image acquisition system (CaseViewer2.4, 3DHISTECH, Budapest, Hungary).

2.3. Cell Isolation Culture

Several chicken embryo glandular stomachs were isolated under sterile condition, and 0.25% trypsin was digested at 37 °C for 10 min. Digestion was terminated with DMEM (HyCyte, Suzhou, Jiangsu, China) containing 10% fetal bovine serum (FBS, HyCyte, Suzhou, Jiangsu, China). Use 0.1% type II collagenase for digestion for another 10 min. Then, DMEM containing 10% FBS was added to terminate digestion. Finally, the cells were resuspended in DMEM containing 10% FBS and uniformly inoculated into T25 cell vials. The new paper was placed horizontally in an incubator at 37 °C and 5%CO_2 for culture. Then, 0.25% trypsin and 0.1% type II collagenase were purchased from Shanghai Biyuntian Biotechnology Co., Ltd. (Shanghai, China). Control group (Con): the glandular gastric epithelial cells were cultured normally. ZEA group: the cells were cultured with 2 μM of ZEA. ZEA and low-dose GA co-treatment group (ZEA+LGA): the cells were cultured with 2 μM ZEA and 50 μM GA. ZEA and medium-dose GA co-treatment group (ZEA+MGA): the cells were cultured with 2 μM of ZEA and 100 μM of GA. ZEA and high-dose GA co-treatment group (ZEA+HGA): the cells were cultured with 2 μM of ZEA and 150 μM of GA.

2.4. TUNEL Analysis

The paraffin sections were dewaxed twice for 10 min, before and after, and then hydrated with a high-to-low concentration of ethanol solution. ProteinaseK working fluid was used to permeate sections to promote TUNEL working fluid penetration into the cells [23]. Finally, TUNEL working solution was used to observe the apoptosis of cells in the tissues under a fluorescence microscope.

2.5. Reactive Oxygen Species (ROS) Detection

The cells were cultured in groups, washed with PBS, and diluted. Then, a DCFH-DA probe was added, and the cells were incubated at 37 °C away from light for 30 min and then washed with PBS after discarding the solution. Fluorescence intensity was observed under a fluorescence microscope [24]. The ROS kit was acquired from Nanjing Jiancheng Biotechnology Co. (Nanjing, China).

2.6. Antioxidant Enzyme Detection

A total of 0.1 g of chicken glandular stomach tissue was weighed, 1 mL of normal saline was added, the mixture was ground to a homogenate at 4 °C and centrifuged at 4000 rpm at 4 °C for 15 min, and the supernatant was obtained. A total of 1 mL of PBS-suspended cells was then ground at 4 °C and centrifuged at 4000 rpm for 15 min to obtain a supernatant. MDA content was determined via the TBA method, SOD activity was determined via the hydroxylamine method, CAT activity was determined via the ammonium molybdate method, T-AOC activity was determined via the colorimetric method, and GSH-PX activity was determined via the enzymatic reaction method. All procedures followed the instructions of the operations conducted at Nanjing Jiancheng Co., Ltd., Nanjing, China.

2.7. ELISA

An ELISA kit (Meimian, Yancheng, Jiangsu, China) was used to detect the production of inflammatory factors IL-6, IL-10, IL-1β, and TNF-α in chicken glandular stomach tissue; 0.1g of tissue was weighed and ground with 1 mL of pre-cooled PBS; cells were collected and suspended with PBS; and the tissue and cells were thoroughly ground with a Tissue Lyser. Centrifugation took place at 4000 rpm for 15 min. Then, the kit instructions were followed for the test. OD values were measured at 450 nm using an enzyme-labeled instrument. The protein concentration of each factor in the sample was calculated by drawing a standard curve.

2.8. qRT-PCR

The mRNA expression level was detected via qRT-PCR. RNA from chicken glandular stomach tissues and cells was extracted using Trizol (Invitrogen, Carlsbad, CA, USA), and reverse transcription was conducted with the first-strand cDNA synthesis kit (Bioer Technology, Hangzhou, China). The target genes were amplified with a 20 μL system (1 μL of cDNA, 10 μL of 2×SYBR Green PCR premix, 3.4 μL of water without RNase, 0.4 μL of a forward primer, and 0.4 μL of a reverse primer), and the total number of cycles was 40. Subsequently, qPCR analysis was conducted on a Light Cycler@480 System (Roche, Basel, Switzerland) following the BioEasy Master Mix kit instructions. With GAPDH as the internal reference gene, the calculated result is expressed as $2^{-\Delta\Delta Ct}$, and the primers used in this experiment are shown in Table 1.

Table 1. Primer sequences for real-time PCR amplification.

Name	Primer Sequence Forward (5′-3′)	Primer Sequence Reverse (5′-3′)
Caspase7	AAATGAACACGGAAAACAAC	TTCTGTCTGGCTTTGCATC
Caspase9	CTCGTGGTCATCCTCTCCCAT	CCTCTCAAACTCGGGCACTGG
Bcl2	GAGTTCGGCGGCGTGATGTG	CTCGGTCATCCAGGTGGCAATG
Bax	ACTCTGCTGCTGCTCTCCTCTC	ATCCACGCAGTGCCAGATGTAATC
Caspase3	AGCAGACAGTGGACCAGATGAAAC	GGCGTGTTCCTTCAGCATCCTAC
Caspase1	GGGCTGGCAGGACTCTGATAGG	TCTCTCCCGTGGCTGGTATATGTC
NLRP3	GCTCCTTGCGTGCTCTAAGACC	TTGTGCTTCCAGATGCCGTCAG

Table 1. *Cont.*

Name	Primer Sequence Forward (5′-3′)	Primer Sequence Reverse (5′-3′)
ASC	CCAGCAAAGCCTCCGCCAAG	CAGCAATCCAGCCACCTCATACG
NFκB	GCTCACAAAGGCAGTCTCACCAG	AGGTCTCTACGCCGCTGTCAC
IκB-α	TGGCGATCATTCACGAGGAAAAGG	GTCTGGCTGAGGTTGTTCTGGAAG
RIPK1	GCCTGTTGAGAGCTGTGAGACTTC	AGGGGTGTATGGTGGAACTACTGG
RIPK3	CCCTCCATTCACTCATCCACAAGC	GCAGTGGCAGTCAGCAGGAAG
MLKL	TGCGAGCACCGATACCAGGAG	TCCAGGTCAGCCAGCAAGTCC
GAPDH	CAGAACATCATCCCAGCGTCCAC	CGGCAGGTCAGGTCAACAACAG

Caspase7, Cysteine-aspartic acid protease 7; Caspase9, Cysteine-aspartic acid protease 9; Bcl-2, B-cell lymphoma 2; Bax, Bcl-2-associated X Protein; Caspase3, Cysteine-aspartic acid protease 3; Caspase1, Cysteine-aspartic acid protease 1; NLRP3, Nucleotide-binding oligomerization domain-like receptor protein 3; ASC, Apoptosis-associated speck-like protein containing a CARD; NFκB, Nuclear Factor kappa-light-chain-enhancer of activated B cells; IκB-α, Inhibitor of kappa B-α; RIPK1, Receptor-Interacting Protein Kinase 3; RIPK1, Receptor-Interacting Protein Kinase 3; MLKL, Mixed Lineage Kinase Domain-Like; GAPDH, Glyceraldehyde-3-phosphate dehydrogenase; F, forward; R, reverse.

2.9. Western Blotting

Protein was extracted from glandular stomach tissue and cells and then ground. The concentration of histamine was determined and adjusted following the BCA (Beyotime, Shanghai, China) method. The supernatant was obtained from chicken gastric epithelial cells via ice lysis for 30 min and centrifugation at 4 °C 12,000 rpm for 10 min. The cell protein concentration was measured following the BCA method and then allowed to level out. The protein samples were separated via 12% SDS-PAGE and transferred to an NC membrane (Sigma, Mannheim, Germany). After 5% skim milk was sealed for 2 h, the NC film was placed in the primary antibody and incubated at 4 °C overnight. After TBST cleaning, the membrane was incubated in the secondary antibody for 1 h away from light. The image was exposed and retained with ECL luminescent solution (Gibco, Baltimore, ML, USA) and analyzed with ImageJ (National Institutes of Health, Bethesda, MD, USA) [25].

2.10. AO/EB

The apoptosis and necrosis of chicken glandular gastric epithelial cells were detected via the AO/EB method. AO and EB were added 1:1 to 1mL of PBS, and 200 μL were added to each well of the cell plate and stained in the dark for 10 min. Images were obtained by fluorescence microscopy [26].

2.11. Statistical Analysis

All data are presented as standard deviation (SD) ± mean, and at least three biological replicates were performed for all experiments. Group differences were analyzed by one-way analysis of variance (ANOVA) using GraphPad Prism software 9.0. "*" indicates that a p-value < 0.05 is considered significant.

3. Results

3.1. GA Alleviates ZEA-Induced Tissue Damage and Oxidative Stress

A histological examination with HE staining (Figure 1A) showed that the glandular gastric villi were arranged regularly and that the glandular gastric structure was complete in the Con group. In contrast, the ZEA group showed villi breakage and the presence of many inflammatory cells. GA gradually restored villi arrangement and structural integrity in a dose-dependent manner, but a small amount of inflammatory infiltration and villi breakage still occurred. In order to investigate whether GA can participate in oxidative stress and reduce the effect of ZEA on chicken glandular stomach, ROS and antioxidant enzyme activities in chicken glandular stomach were detected in this experiment. ZEA caused a significant increase in ROS levels, while the ZEA-induced ROS levels decreased after GA intervention ($p < 0.001$) (Figure 1B). MDA is a marker of lipid peroxidation, and the production of MDA significantly increased under the effects of ZEA, while the level of MDA significantly decreased under the effects of GA ($p < 0.001$) (Figure 1C). With the increase in GA dose, the activities of CAT, SOD, GSH-PX, and T-AOC increased significantly, while ZEA decreased the activities of the above antioxidant enzymes (Figure 1D–G).

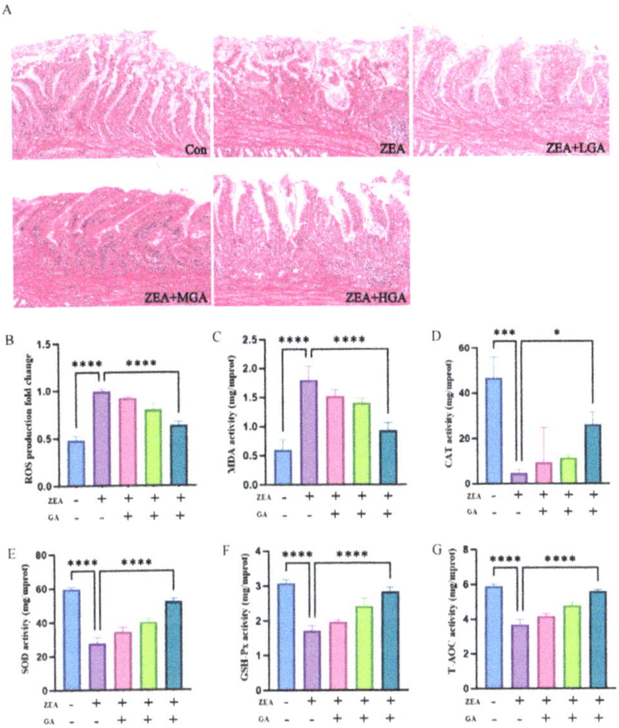

Figure 1. GA alleviates the effects of histological damage and oxidative stress caused by ZEA. (**A**) HE staining of chicken glandular stomach sections in the control group and at different GA concentrations. (**B**–**G**) Effects of control group and different concentrations of GA on oxidative stress in glandular gastric tissue of chickens ($n = 10$). * Means with different superscripts within the same column differ ($p < 0.05$). *** Means with different superscripts within the same column differ ($p < 0.005$). **** Means with different superscripts within the same column differ ($p < 0.001$).

3.2. *GA Relieved ZEA-Induced Interstitial Inflammation*

In order to investigate the mechanism of the inflammatory response induced by ZEA in the chicken glandular stomach and the alleviating effect of GA, the expression of the inflammatory cytokine *IL-6*, *IL-10*, *IL-1β*, and *TNF-α* genes was detected (Figure 2A). After exposure to ZEA, the mRNA levels of *IL-6*, *IL-1β*, and *TNF-α* increased, while the expression of *IL-10* mRNA decreased. Furthermore, the mRNA levels of inflammatory factors *IL-6*, *IL-1β*, and *TNF-α* decreased and the mRNA levels of *IL-10* increased after the application of GA. At the protein level, the expression of inflammatory cytokines was consistent with the trend for their mRNA levels. In the ZEA group, the expression of IL-6, IL-1β, and TNF-α significantly increased, while the expression of IL-10 protein decreased. After GA intervention, the expression of IL-6, IL-1β, and TNF-α significantly decreased, while the expression of IL-10 significantly increased ($p < 0.001$) (Figure 2B–E). In order to further examine whether the NFκB/IκB-α-signaling pathway participates in the mechanism behind how GA alleviates glandular gastric inflammation in chickens, the expression of the signaling pathway genes and proteins was detected. ZEA exposure significantly increased the mRNA levels of *NLRP3*, *ASC*, *Caspase1*, *NFκB p65*, and *IκB-α*. However, the mRNA levels of *NLRP3*, *ASC*, *Caspase1*, *NFκB p65*, and *IκB-α* decreased when GA was added to the feed (Figure 2F). The protein expression of the NFκB/IκB-α-signaling pathway was similar to its mRNA expression. Compared with the ZEA exposure group, the group in which GA was added significantly reduced the protein expression of NLRP3, ASC, Caspase1, p-NFκB p65, and IκB-α ($p < 0.05$) (Figure 2H–L).

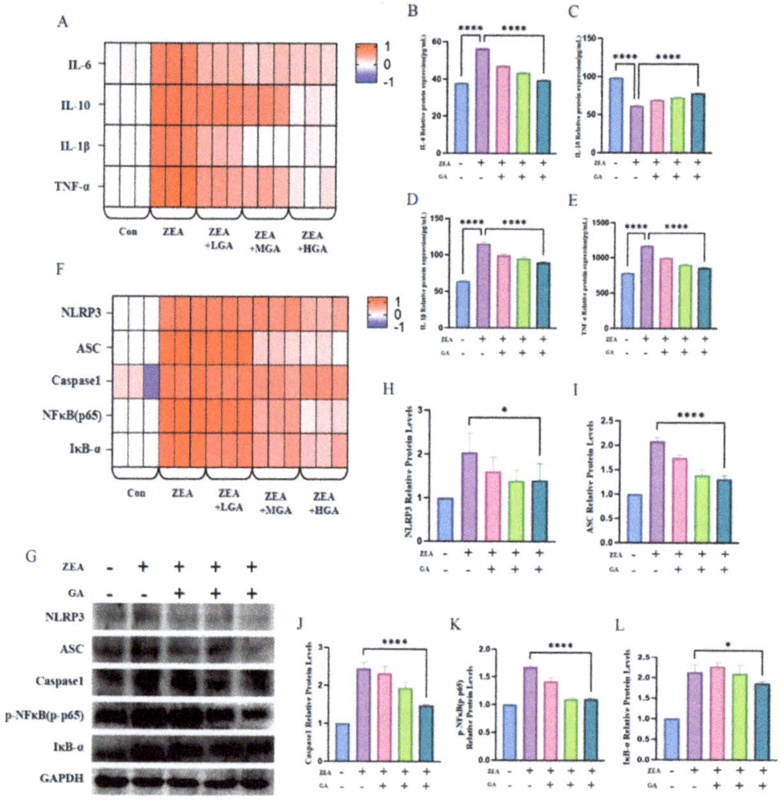

Figure 2. Effects of ZEA exposure and GA on glandular gastric inflammation and the NFκB/IκB−α−signaling pathway in chickens. (**A**) mRNA levels of *IL-6*, *IL−10*, *IL−1β*, and *TNF-α* ($n = 3$). (**B**) Protein expression of IL-6 ($n = 10$). (**C**) Expression of IL−10 protein ($n = 10$). (**D**) Protein expression of IL−1β ($n = 10$). (**E**) Protein expression of TNF-α ($n = 10$). (**F**) mRNA levels of *NLRP3*, *ASC*, *Caspase1*, *NFκB p65*, and *IκB−α* ($n = 3$). (**G**) Western blot images of proteins associated with the NFκB/IκB−α−signaling pathway ($n = 10$). (**H**) NLRP3 protein expression ($n = 10$). (**I**) Protein expression of ASC ($n = 10$). (**J**) Protein expression of Caspase1 ($n = 10$). (**K**) Protein expression of p−NFκB ($n = 10$). (**L**) Protein expression of IκB−α ($n = 10$). * Means with different superscripts within the same column differ ($p < 0.05$). **** Means with different superscripts within the same column differ ($p < 0.001$).

3.3. GA Reduced ZEA-Induced Histiocytic Death

TUNEL staining was used to detect cell apoptosis in the gizzard tissue of chickens. Compared with the control group, the relative fluorescence intensity of TUNEL-positive cells in the ZEA group significantly increased. Compared with the ZEA group, the relative fluorescence intensity of TUNEL-positive cells in the GA group was lower and dose-dependent (Figure 3A). Further detection of the apoptosis-related factors showed that ZEA significantly promoted the mRNA expression of pro-apoptotic factors *Caspase7*, *Caspase9*, *Bax*, and *Caspase3* and inhibited the mRNA expression of anti-apoptotic factor *Bcl2*. GA decreased the mRNA expression of pro-apoptotic factors *Caspase7*, *Caspase9*, *Bax*, and *Caspase3* and increased the mRNA expression of anti-apoptotic factor *Bcl2* (Figure 3B). The expression levels of the apoptosis-related factor proteins were basically consistent with their mRNA expression levels (Figure 3C–H). In addition, ZEA exposure increased the mRNA expression levels of cell necrosis regulators *RIPK1*, *RIPK3*, and *MLKL*, while the use of GA decreased the mRNA expression levels of *RIPK1*, *RIPK3*, and *MLKL* (Figure 3L). Furthermore, the expression of the p-MLKL, p-RIPK1, and p-RIPK3 proteins was detected (Figure 3J). Notably, the phosphorylated expression of RIPK1, RIPK3, and MLKL plays an important role in this process. A phosphorylated protein analysis showed that ZEA significantly upregulated the expression levels of p-RIPK1, p-RIPK3, and p-MLKL, and the addition of GA in the diet decreased the expression levels of p-RIPK1, p-RIPK3, and p-MLKL ($p < 0.001$) (Figure 3K–M).

3.4. GA Inhibits ZEA-Induced Cellular Oxidative Stress

In order to investigate whether GA inhibits oxidative stress and protects chicken glandular gastric epithelial cells, primary cells were first extracted from chicken embryos for passage culture (Figure 4A). Fluorescence was used to detect the expression of reactive oxygen species (ROS) in chicken glandular gastric epithelial cells. Compared with the control group, the ROS level significantly increased after ZEA treatment, and the level of ROS decreased with GA intervention (Figure 4B). MDA is a marker of lipid peroxidation. Under the action of ZEA, the production of MDA significantly increased, while under the action of GA, the level of MDA significantly decreased ($p < 0.001$) (Figure 4C). Furthermore, ZEA decreased the levels of catalase (CAT), total superoxide dismutase (SOD), and GAutathione peroxidase (GSH-PX) and the total antioxidant capacity (T-AOC), and the intervention of GA increased the expression levels of CAT, SOD, GSH-PX, and T-AOC in a dose-dependent way ($p < 0.05$) (Figure 4D–H).

3.5. GA Protects Cells by Alleviating ZEA-Induced Inflammation Through the Nκb Pathway

In order to investigate the mechanism of inflammation induced by ZEA in chicken glandular stomach epithelial cells and the alleviating effect of GA, the expression of the inflammatory cytokine *IL-6*, *IL-10*, *IL-1β*, and *TNF-α* genes was detected (Figure 5A). After cell exposure to ZEA, the *IL-6*, *IL-1β*, and *TNF-α* mRNA levels increased, and *IL-10* mRNA expression decreased. After the addition of GA, the mRNA levels of inflammatory factors *IL-6*, *IL-1β*, and *TNF-α* decreased, and the expression level of *IL-10* increased. The protein results obtained via ELISA were consistent with the mRNA results (Figure 5B–E). To further verify the role of the NFκB/IκB-α-signaling pathway in the cells, the expression of related genes and proteins was detected. ZEA exposure significantly increased the mRNA levels of *NLRP3*, *ASC*, *Caspase1*, *NFκB p65*, and *IκB-α*. After the addition of GA, the mRNA levels of *NLRP3*, *ASC*, *Caspase1*, *NFκB p65*, and *IκB-α* were reduced (Figure 5F). The protein expression of the NFκB/IκB-α-signaling pathway was similar to its mRNA expression. Compared with the ZEA exposure group, the group with the addition of GA significantly reduced the protein expression of NLRP3, ASC, Caspase1, p-NFκB p65, and IκB-α ($p < 0.01$) (Figure 5G–L).

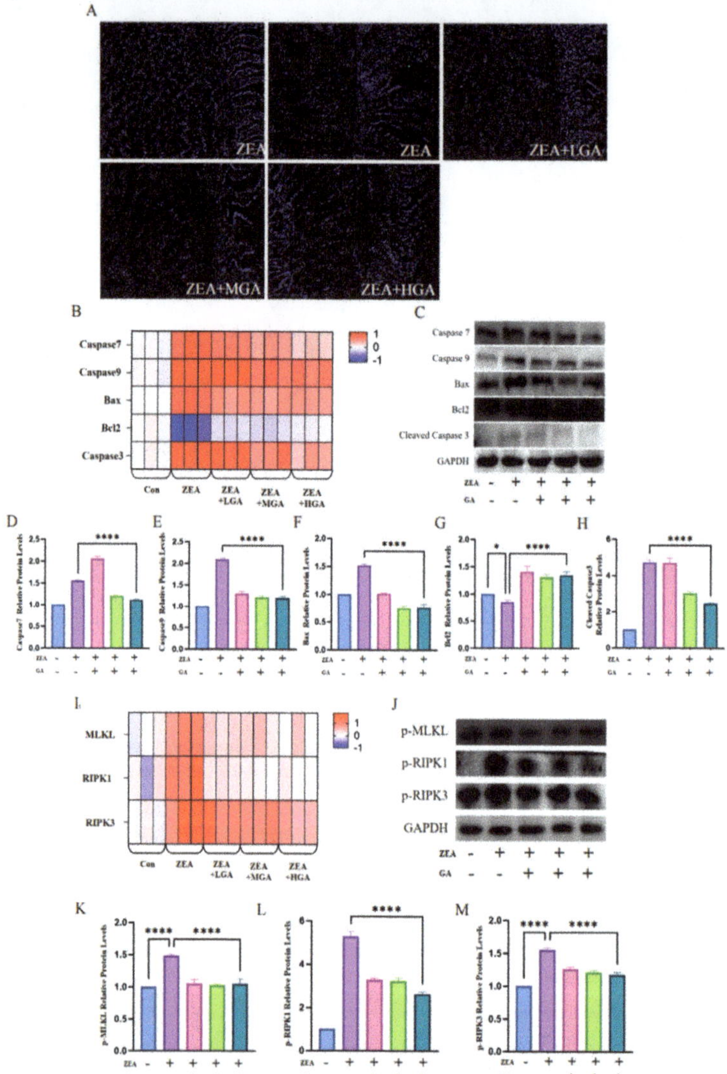

Figure 3. Effect of GA intervention on ZEA−induced cell death. (**A**) TUNEL−positive cells are stained green. Scale bar = 50. (**B**) mRNA levels of *Caspase7*, *Caspase9*, *Bax*, *Bcl2*, and *Caspase3* ($n = 3$). (**C**) Protein immunoblot images of Caspase7, Caspase9, Bax, Bcl2, and Caspase3 ($n = 10$). (**D**) Protein expression of Caspase7 ($n = 10$). (**E**) Protein expression of Caspase9 ($n = 10$). (**F**) Protein expression of Bax ($n = 10$). (**G**) Protein expression of Bcl2 ($n = 10$). (**H**) Cleaved Caspase3 protein expression ($n = 10$). (**I**) mRNA levels of *RIPK1*, *RIPK3*, and *MLKL* ($n = 3$). (**J**) Western blot images of p−MLKL, p−RIPK1, and p−RIPK3 ($n = 10$). (**K**) Protein expression of p−MLKL ($n = 10$). (**L**) Expression of p−RIPK1 protein ($n = 10$). (**M**) Protein expression of p−RIPK3 ($n = 10$). * Means with different superscripts within the same column differ ($p < 0.05$). **** Means with different superscripts within the same column differ ($p < 0.001$).

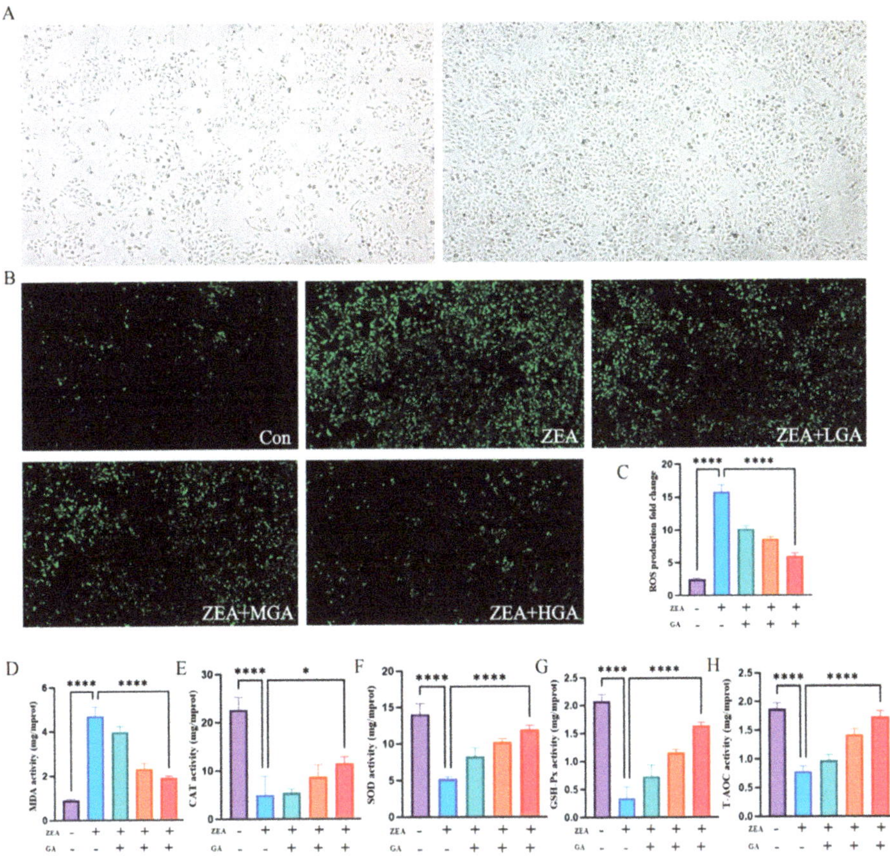

Figure 4. Effects of GA on ZEA−induced oxidative stress on chicken gastric gland epithelial cells. (**A**) Isolation and culture of chicken glandular gastric epithelial cells. (**B**) ROS staining of chicken glandular gastric epithelial cells under different concentrations of GA. (**C**–**H**) Effects of control group and different concentrations of GA on oxidative stress of chicken glandular gastric epithelial cells (n = 10). * Means with different superscripts within the same column differ ($p < 0.05$). **** Means with different superscripts within the same column differ ($p < 0.001$).

3.6. GA Protects Cells and Reduces ZEA-Induced Cell Death

To verify that GA can reduce ZEA-induced apoptosis in chicken glandular gastric epithelial cells, we examined the expression levels of the *Caspase7*, *Caspase9*, *Bax*, *Bcl2*, and *Cleaved Caspase3* genes. The results showed that ZEA exposure significantly increased the expression of *Caspase7*, *Caspase9*, *Bax*, and *Cleaved Caspase3* and decreased the expression of the *Bcl2* gene (Figure 6A). At the same time, the expression results were verified at the protein level, and the results were identical to those obtained for their respective genes. ZEA exposure upregulated the protein levels of Caspase7, Caspase9, Bax, Bcl2, and Cleaved Caspase3. GA intervention resulted in dose-dependent decreases in the expression of the Caspase7, Caspase9, Bax, Bcl2, and Cleaved Caspase3 proteins ($p < 0.005$) (Figure 6B–G). Then, AOEB staining was performed to detect the number of live and dead cells within the cells. As shown in the figure, the number of dead cells increased significantly after ZEA exposure, and GA could effectively alleviate cell death caused by ZEA (Figure 6H).

Subsequently, we tested the expression levels of the *MLKL, RIPK1,* and *RIPK3* genes, and the results showed that ZEA exposure significantly increased the expression levels of the *MLKL, RIPK1,* and *RIPK3* genes, but they also showed that, after the addition of GA, the expression levels of the *MLKL, RIPK1,* and *RIPK3* genes were significantly reduced (Figure 6I). To further verify the above results, a Western blot was conducted to detect the expression levels of p-MLKL, p-RIPK1, and p-RIPK3. At the protein level, the expression level of necrosis factor protein was basically consistent with the mRNA expression level (Figure 6J–M).

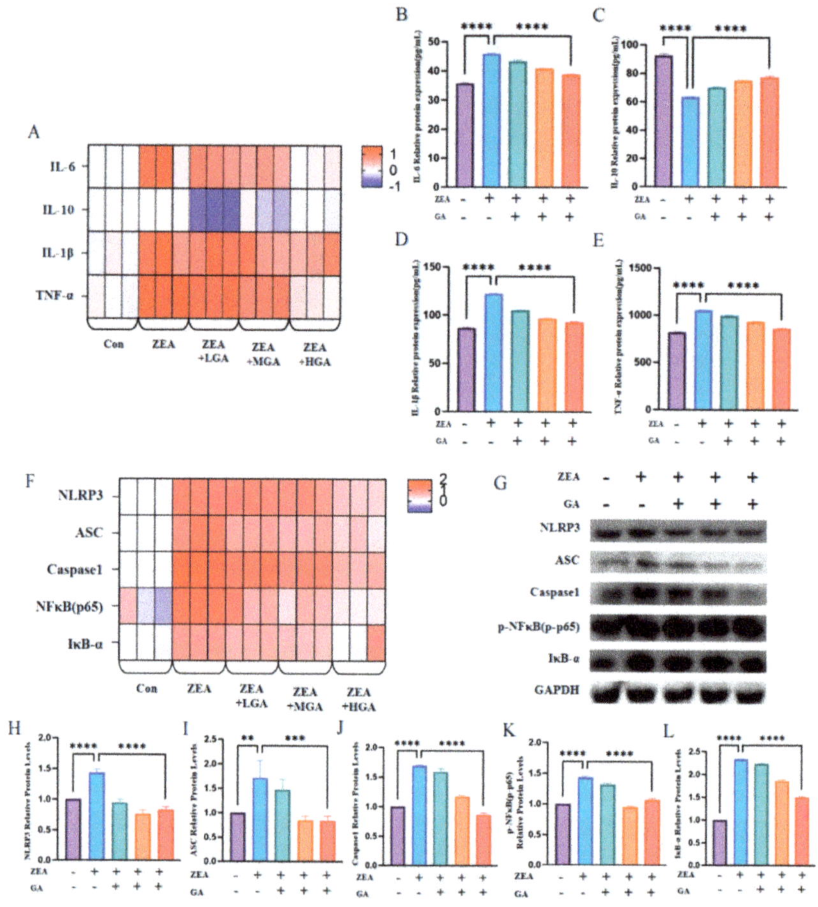

Figure 5. Effects of ZEA exposure and GA on inflammation and NFκB/IκB−α−signaling pathways in chicken Glandular gastric epithelial cells. (**A**) mRNA levels of *IL-6, IL-10, IL-1β,* and *TNF-α* (n = 3). (**B**) Protein expression of IL-6 (n = 10). (**C**) Expression of IL−10 protein (n = 10). (**D**) Protein expression of IL−1β (n = 10). (**E**) Protein expression of TNF-α (n = 10). (**F**) mRNA levels of *NLRP3, ASC, Caspase1, NFκB p65,* and *IκB−α* (n = 3). (**G**) Western blot images of proteins associated with the NFκB/IκB−α−signaling pathway (n = 10). (**H**) NLRP3 protein expression (n = 10). (**I**) Protein expression of ASC (n = 10). (**J**) Protein expression of Caspase1 (n = 10). (**K**) Protein expression of p−NFκB (n = 10). (**L**) Protein expression of IκB−α (n = 10). ** Means with different superscripts within the same column differ ($p < 0.01$). *** Means with different superscripts within the same column differ ($p < 0.005$). **** Means with different superscripts within the same column differ ($p < 0.001$).

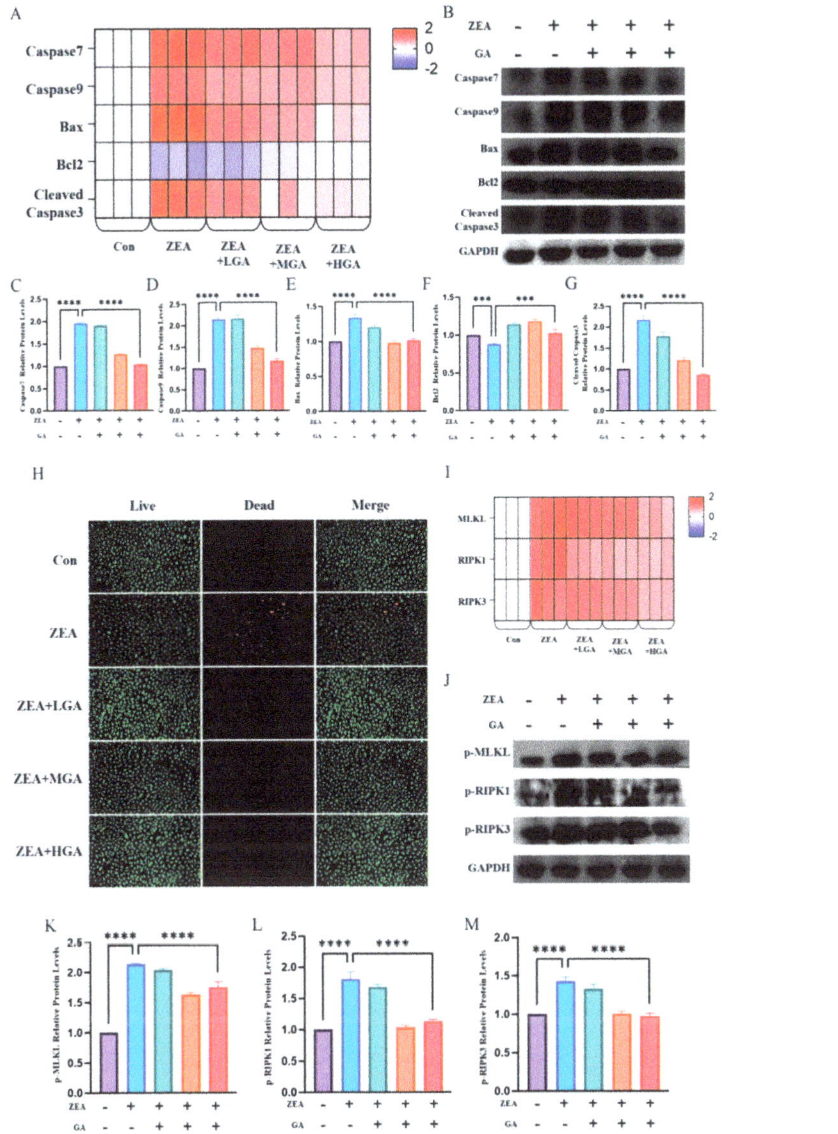

Figure 6. Effect of GA on ZEA−induced glandular gastric epithelial cell death in chickens. (**A**) mRNA levels of *Caspase7*, *Caspase9*, *Bax*, *Bcl2*, and *Caspase3* (n = 3). (**B**) Western blot images of Caspase7, Caspase9, Bax, Bcl2, and Caspase3 proteins (n = 10). (**C**) Protein expression of Caspase7 (n = 10). (**D**) Protein expression of Caspase9 (n = 10). (**E**) Protein expression of Bax (n = 10). (**F**) Protein expression of Bcl2 (n = 10). (**G**) Cleaved Caspase3 protein expression (n = 10). (**H**) A: AO/EB staining of control cells exposed to ZEA, ZEA+LGA, ZEA+MGA, and ZEA+HGA. Living cells are bright green, apoptotic cells are bright orange, and dead cells are red. (**I**) mRNA levels of *MLKL*, *RIPK1*, and *RIPK3* (n = 3). (**J**) Western blot images of p−MLKL, p−RIPK1, and p−RIPK3 proteins (n = 10). (**K**) Protein expression of p−MLKL (n = 10). (**L**) Expression of p−RIPK1 protein (n = 10). (**M**) Protein expression of p−RIPK3 (n = 10). *** Means with different superscripts within the same column differ ($p < 0.005$). **** Means with different superscripts within the same column differ ($p < 0.001$).

4. Discussion

ZEA is a mycotoxin that is widely found in moldy grains and can cause tissue damage in the digestive, particularly the liver, and reproductive systems. ZEA is relatively well tolerated in poultry, but long-term consumption can affect the growth performance of chicks, impair liver and kidney function, and reduce antioxidant performance [27,28]. GA has a variety of biological properties, including antioxidant, antifungal, and anti-inflammatory properties, and it has become a popular therapeutic drug [29,30]. The results show that dietary ZEA can induce inflammation of gastric gland tissue and lead to necrosis and apoptosis of gastric gland epithelial cells. GA from glycyrrhiza extract can alleviate the glandular gastric injury induced by ZEA. In this study, the glandular gastric tissue morphology in the tissue slices after adding GA to the diet tended to be normal, and the number of lesions was significantly reduced compared with that in the ZEA group. Therefore, GA may, to some extent, alleviate tissue damage caused by ZEA ingestion.

The antioxidant system plays an important role in the growth and development of birds. The content of MDA reflects the antioxidant capacity of the body and the lipid peroxide damage in tissues. ZEA exposure has been shown to significantly increase MDA content [31]. At the same time, the expression level of other antioxidant enzymes decreased, which caused oxidative stress injury to the chicken glandular stomach. Studies have confirmed that the addition of glycyrrhiza extract in feed can effectively improve the total antioxidant capacity and SOD activity of broilers [32]. Adding licorice extract to drinking water can increase the activities of GSH-PX and catalase and decrease the expression of MDA in broilers. GA is a natural antioxidant that can remove excess ROS and increase the activity of antioxidant enzymes [33]. In this study, adding different doses of GA to the feed supplemented with ZEA can significantly improve the antioxidant capacity of the chicken glandular stomach and reduce the level of MDA. In conclusion, GA can effectively reduce the damage caused by ZEA on the antioxidant function of chicken glandular stomach.

When inflammation occurs in the body, it is inevitably accompanied by the release of pro-inflammatory factors and the inhibition of anti-inflammatory factors. ZEA induces glandular gastric inflammation in chickens, increases the expression of pro-inflammatory factors (IL-6, IL-1β, and TNF-α), and inhibits the expression of anti-inflammatory factors (IL-10) [34]. Dietary ZEA supplementation can significantly increase the expression of IL-6, IL-1β, and TNF-α in glandular gastric tissue of chickens and decrease the expression of IL-10, suggesting that ZEA can promote tissue inflammation. GA reduces the binding of ZEA to organs by increasing the body's antioxidant capacity. In this study, we found that the expression of IL-6, IL-1β, and TNF-α decreased and the expression of IL-10 increased after GA was added to ZEA-exposed diets, suggesting that GA can reduce ZEA-induced inflammation. Activation of the NFκB-signaling pathway can promote the expression of inflammatory cytokines [35]. In our study, the levels of NLRP3, ASC, Caspase1, p-NFκB, and IκB-α were significantly increased in the ZEA exposure group. During inflammation, Caspase1 is activated through autocatalytic activation [36]. The NLRP3 inflammasome is critical in host immunity to fungal, bacterial, and viral infections [37–40]. The pyrin domain of NLRP3 interacts with the pyrin domain of ASC to initiate inflammasome assembly. In addition, the NFκB-signaling pathway corresponds to NLRP3 induction [41]. The role of ROS in NLRP3 activation is still widely debated. ROS is considered to be a common signal during NLRP3 inflammasome activation [42,43]. Recent studies have shown that ROS plays a tissue-specific role in the activation of the NLRP3 inflammasome [44]. Therefore, the NFκB-signaling pathway is closely related to inflammation and apoptosis.

Inflammation and oxidative stress can induce programmed apoptosis and necrosis, and previous studies have confirmed that ZEA can also induce apoptosis and necrosis of human hepatocytes, HepaRG. Therefore, in this study, we measured and analyzed the indicators related to apoptosis and necrosis. The results showed that ZEA exposure significantly inhibited the expression of Bcl2, a classical anti-apoptotic molecule that inhibits cell death by preventing mitochondrial membrane penetration. Studies have shown that NFκB targeting Bcl2 regulates cell death, and a decrease in Bcl2 with an increase in Bax indicates an increase in cell apoptosis. In our results, ZEA exposure increased the expression level of Bax and apoptosis-related factors such as Caspase9, both confirming that ZEA induces apoptosis through the NFκB-signaling pathway. In this study, we found that ZEA can also activate the RIPK1/RIPK3/MLKL-signaling pathway to promote cell necrosis, and the in vitro results showed that the level of cell necrosis induced through exposure was significantly higher than that of apoptosis.

Our results suggest that exposure to ZEA activates the NFκB/IκB-α pathway, leading to tissue inflammation and increased expression of pro-inflammatory factors. Activation of the NFκB/IκB-α pathway decreased the expression level of Bcl2 and increased the expression level of Bax, inducing cell apoptosis. In addition, ZEA activates the RIPK1/RIPK3/MLKL-signaling pathway, inducing the occurrence of necrosis. However, the number of instances of ZEA exposure-induced cell necrosis was significantly higher than that of apoptosis, suggesting that ZEA causes tissue and cell damage primarily through cell necrosis. These results indicate that ZEA exposure can damage the glandular stomach of chickens and affect the health of poultry. Studies have shown that the body cannot completely metabolize ZEA after ingestion, and some ZEA remains in the body, causing sustained damage [45]. In recent years, research on GA has made remarkable progress regarding the treatment of inflammatory tissue damage via antioxidative stress. As a natural antioxidant, GA may become a key drug in the treatment and control of toxic diseases in chickens, so we hope that our study can provide some basis for the treatment of avian diseases with GA.

5. Conclusions

Our study found that ZEA exposure causes glandular gastric inflammation, apoptosis, and programmed necrosis in chickens. ZEA exposure can induce oxidative stress, significantly increase the levels of oxidative stress markers ROS and MDA, and inhibit the activity of other antioxidant enzymes. In this study, ZEA exposure induced apoptosis and necrosis of cells. The therapeutic effect of GA showed a dose-dependent pattern. These results demonstrated that GA mitigated the toxicological effects of ZEA in chicken glandular stomach and chicken glandular stomach epithelial cells. This study enriches the toxicological research on ZEA and the therapeutic effect of GA, which is of great significance for the protection of grain feed and poultry health.

Author Contributions: Conceptualization, T.S. and F.W.; methodology, M.Q.; software, J.W.; validation, F.W., M.Q. and J.W.; formal analysis, F.W.; investigation, M.Q.; resources, T.S.; data curation, T.S.; writing—original draft preparation, T.S.; writing—review and editing, M.G.; visualization, F.W.; supervision, M.G.; project administration, M.G.; funding acquisition, M.G. All authors have read and agreed to the published version of the manuscript.

Funding: This work was supported by the Natural Science Foundation of Heilongjiang Province (YQ2022C019).

Institutional Review Board Statement: This study was approved by the Institutional Animal Care and Use Committee of Northeast Agriculture University. The approval number was NEAU2023046, and the date was 1 March 2023.

Informed Consent Statement: The authors declare that they have no known competing financial interests or personal relationships that could have appeared to influence the work reported in this paper.

Data Availability Statement: The original contributions presented in the study are included in the article, further inquiries can be directed to the corresponding author.

Conflicts of Interest: The authors declare no conflicts of interest.

References

1. Ropejko, K.; Twarużek, M. Zearalenone and Its Metabolites-General Overview, Occurrence, and Toxicity. *Toxins* **2021**, *13*, 35. [CrossRef] [PubMed]
2. Schothorst, R.C.; van Egmond, H.P. Report from SCOOP task 3.2.10 "collection of occurrence data of Fusarium toxins in food and assessment of dietary intake by the population of EU member states". Subtask: Trichothecenes. *Toxicol. Lett.* **2004**, *153*, 133–143. [CrossRef]
3. Han, X.; Huangfu, B.; Xu, T.; Xu, W.; Asakiya, C.; Huang, K.; He, X. Research Progress of Safety of Zearalenone: A Review. *Toxins* **2022**, *14*, 386. [CrossRef]
4. Raj, J.; Farkaš, H.; Jakovčević, Z.; Medina, A.; Magan, N.; Čepela, R.; Vasiljević, M. Comparison of multiple mycotoxins in harvested maize samples in three years (2018-2020) in four continents. *Food Addit. Contam. Part A* **2022**, *39*, 599–608. [CrossRef] [PubMed]
5. Zhu, F.; Zhu, L.; Xu, J.; Wang, Y.; Wang, Y. Effects of moldy corn on the performance, antioxidant capacity, immune function, metabolism and residues of mycotoxins in eggs, muscle, and edible viscera of laying hens. *Poult. Sci.* **2023**, *102*, 102502. [CrossRef]
6. Bi, Z.; Gu, X.; Xiao, Y.; Zhou, Y.; Bao, W.; Wu, S.; Wang, H. Analysis of the Roles of the ISLR2 Gene in Regulating the Toxicity of Zearalenone Exposure in Porcine Intestinal Epithelial Cells. *Toxins* **2022**, *14*, 639. [CrossRef]
7. Shen, T.; Miao, Y.; Ding, C.; Fan, W.; Liu, S.; Lv, Y.; Gao, X.; De Boevre, M.; Yan, L.; Okoth, S.; et al. Activation of the p38/MAPK pathway regulates autophagy in response to the CYPOR-dependent oxidative stress induced by zearalenone in porcine intestinal epithelial cells. *Food Chem. Toxicol.* **2019**, *131*, 110527. [CrossRef]
8. Knapp BK, Bauer LL, Swanson KS, Tappenden KA, Fahey GC Jr, de Godoy MR. Soluble fiber dextrin and soluble corn fiber supplementation modify indices of health in cecum and colon of Sprague-Dawley rats. *Nutrients* **2013**, *5*, 396–410. [CrossRef] [PubMed]
9. Debevere, S.; Cools, A.; De Baere, S.; Haesaert, G.; Rychlik, M.; Croubels, S.; Fievez, V. In Vitro Rumen Simulations Show a Reduced Disappearance of Deoxynivalenol, Nivalenol and Enniatin B at Conditions of Rumen Acidosis and Lower Microbial Activity. *Toxins* **2020**, *12*, 101. [CrossRef] [PubMed]
10. Kemboi, D.C.; Antonissen, G.; Ochieng, P.E.; Croubels, S.; Okoth, S.; Kangethe, E.K.; Faas, J.; Lindahl, J.F.; Gathumbi, J.K. A Review of the Impact of Mycotoxins on Dairy Cattle Health: Challenges for Food Safety and Dairy Production in Sub-Saharan Africa. *Toxins* **2020**, *12*, 222. [CrossRef] [PubMed]
11. Tardieu, D.; Travel, A.; Le Bourhis, C.; Metayer, J.-P.; Mika, A.; Cleva, D.; Boissieu, C.; Guerre, P. Fumonisins and zearalenone fed at low levels can persist several days in the liver of turkeys and broiler chickens after exposure to the contaminated diet was stopped. *Food Chem. Toxicol.* **2021**, *148*, 111968. [CrossRef]
12. Fu, Y.; Jin, Y.; Tian, Y.; Yu, H.; Wang, R.; Qi, H.; Feng, B.; Zhang, J. Zearalenone Promotes LPS-Induced Oxidative Stress, Endoplasmic Reticulum Stress, and Accelerates Bovine Mammary Epithelial Cell Apoptosis. *Int. J. Mol. Sci.* **2022**, *23*, 10925. [CrossRef]
13. Yi, Y.; Gao, K.; Zhang, L.; Lin, P.; Wang, A.; Jin, Y. Zearalenone Induces MLKL-Dependent Necroptosis in Goat Endometrial Stromal Cells via the Calcium Overload/ROS Pathway. *Int. J. Mol. Sci.* **2022**, *23*, 10170. [CrossRef] [PubMed]
14. Zhang, P.; Jing, C.; Liang, M.; Jiang, S.; Huang, L.; Jiao, N.; Li, Y.; Yang, W. Zearalenone Exposure Triggered Cecal Physical Barrier Injury through the TGF-β1/Smads Signaling Pathway in Weaned Piglets. *Toxins* **2021**, *13*, 902. [CrossRef] [PubMed]
15. Fan, W.; Shen, T.; Ding, Q.; Lv, Y.; Li, L.; Huang, K.; Yan, L.; Song, S. Zearalenone induces ROS-mediated mitochondrial damage in porcine IPEC-J2 cells. *J. Biochem. Mol. Toxicol.* **2017**, *31*, 10. [CrossRef] [PubMed]
16. Guijarro-Muñoz, I.; Compte, M.; Álvarez-Cienfuegos, A.; Álvarez-Vallina, L.; Sanz, L. Lipopolysaccharide activates Toll-like receptor 4 (TLR4)-mediated NF-κB signaling pathway and proinflammatory response in human pericytes. *J. Biol. Chem.* **2014**, *289*, 2457–2468. [CrossRef]
17. Pistol, G.C.; Gras, M.A.; Marin, D.E.; Israel-Roming, F.; Stancu, M.; Taranu, I. Natural feed contaminant zearalenone decreases the expressions of important pro- and anti-inflammatory mediators and mitogen-activated protein kinase/NF-κB signalling molecules in pigs. *Br. J. Nutr.* **2014**, *111*, 452–464. [CrossRef]

18. Sun, L.; He, D.; Liu, Y.; Wei, Y.; Wang, L. Corynoline protects against zearalenone-induced liver injury by activating the SIRT1/Nrf2 signaling pathway. *J. Biochem. Mol. Toxicol.* **2023**, *37*, e23224. [CrossRef] [PubMed]
19. Xu, J.; Li, S.; Jiang, L.; Gao, X.; Liu, W.; Zhu, X.; Huang, W.; Zhao, H.; Wei, Z.; Wang, K.; et al. Baicalin protects against zearalenone-induced chicks liver and kidney injury by inhibiting expression of oxidative stress, inflammatory cytokines and caspase signaling pathway. *Int. Immunopharmacol.* **2021**, *100*, 108097. [CrossRef]
20. Luo, J.-J.; Zhang, Y.; Sun, H.; Wei, J.-T.; Khalil, M.M.; Wang, Y.-W.; Dai, J.-F.; Zhang, N.-Y.; Qi, D.-S.; Sun, L.-H. The response of glandular gastric transcriptome to T-2 toxin in chicks. *Food Chem. Toxicol.* **2019**, *132*, 110658. [CrossRef]
21. Heidari, S.; Mehri, S.; Hosseinzadeh, H. The genus Glycyrrhiza (Fabaceae family) and its active constituents as protective agents against natural or chemical toxicities. *Phytother. Res.* **2021**, *35*, 6552–6571. [CrossRef] [PubMed]
22. Sharifi-Rad, J.; Quispe, C.; Herrera-Bravo, J.; Belén, L.H.; Kaur, R.; Kregiel, D.; Uprety, Y.; Beyatli, A.; Yeskaliyeva, B.; Kırkın, C.; et al. Glycyrrhiza Genus: Enlightening Phytochemical Components for Pharmacological and Health-Promoting Abilities. *Oxid. Med. Cell Longev.* **2021**, *2021*, 7571132. [CrossRef] [PubMed]
23. Xu, R.; Han, F.X.; Wang, H.R.; Wang, J.J.; Cai, Z.L.; Guo, M.Y. Tea polyphenols alleviate TBBPA-induced inflammation, ferroptosis and apoptosis via TLR4/NF-κB pathway in carp gills. *Fish. Shellfish. Immunol.* **2024**, *146*, 109382. [CrossRef] [PubMed]
24. Cui, J.; Zhu, M.; Sun, X.; Yang, J.; Guo, M. Microplastics induced endoplasmic reticulum stress to format an inflammation and cell death in hepatocytes of carp (Cyprinus carpio). *Aquat. Toxicol.* **2024**, *269*, 106870. [CrossRef]
25. Wu, J.; Zhang, Y.; Liu, T.; Yang, J.; Sun, X.; Gao, X.J. The mechanism of selenium regulating the permeability of vascular endothelial cells through selenoprotein O. *Redox Biol.* **2024**, *70*, 103063. [CrossRef]
26. Zhang, Q.; Wang, F.; Xu, S.; Cui, J.; Li, K.; Shiwen, X.; Guo, M.-Y. Polystyrene microplastics induce myocardial inflammation and cell death via the TLR4/NF-κB pathway in carp. *Fish Shellfish. Immunol.* **2023**, *135*, 108690. [CrossRef] [PubMed]
27. Jia, S.; Ren, C.; Yang, P.; Qi, D. Effects of Intestinal Microorganisms on Metabolism and Toxicity Mitigation of Zearalenone in Broilers. *Animals* **2022**, *12*, 1962. [CrossRef]
28. Król, A.; Pomastowski, P.; Rafińska, K.; Railean-Plugaru, V.; Walczak, J.; Buszewski, B. Microbiology neutralization of zearalenone using Lactococcus lactis and *Bifidobacterium* sp. *Anal. Bioanal. Chem.* **2018**, *410*, 943–952. [CrossRef] [PubMed]
29. Xu, C.; Liang, C.; Sun, W.; Chen, J.; Chen, X. Glycyrrhizic acid ameliorates myocardial ischemic injury by the regulation of inflammation and oxidative state. *Drug Des. Dev. Ther.* **2018**, *12*, 1311–1319. [CrossRef]
30. Wang, H.; Ge, X.; Qu, H.; Wang, N.; Zhou, J.; Xu, W.; Xie, J.; Zhou, Y.; Shi, L.; Qin, Z.; et al. Glycyrrhizic Acid Inhibits Proliferation of Gastric Cancer Cells by Inducing Cell Cycle Arrest and Apoptosis. *Cancer Manag. Res.* **2020**, *12*, 2853–2861. [CrossRef] [PubMed]
31. Xia, S.; Yan, C.; Gu, J.; Yuan, Y.; Zou, H.; Liu, Z.; Bian, J. Resveratrol Alleviates Zearalenone-Induced Intestinal Dysfunction in Mice through the NF-κB/Nrf2/HO-1 Signalling Pathway. *Foods* **2024**, *13*, 1217. [CrossRef]
32. Chen, Y.; Cheng, Y.; Wen, C.; Wang, W.; Kang, Y.; Wang, A.; Zhou, Y. The protective effects of modified palygorskite on the broilers fed a purified zearalenone-contaminated diet. *Poult. Sci.* **2019**, *98*, 3802–3810. [CrossRef]
33. Ojha, S.; Javed, H.; Azimullah, S.; Abul Khair, S.B.; Haque, M.E. Glycyrrhizic acid Attenuates Neuroinflammation and Oxidative Stress in Rotenone Model of Parkinson's Disease. *Neurotox Res.* **2016**, *29*, 275–287. [CrossRef]
34. Lee, P.-Y.; Liu, C.-C.; Wang, S.-C.; Chen, K.-Y.; Lin, T.-C.; Liu, P.-L.; Chiu, C.-C.; Chen, I.-C.; Lai, Y.-H.; Cheng, W.-C.; et al. Mycotoxin Zearalenone Attenuates Innate Immune Responses and Suppresses NLRP3 Inflammasome Activation in LPS-Activated Macrophages. *Toxins* **2021**, *13*, 593. [CrossRef]
35. Guerrero-Hue, M.; García-Caballero, C.; Palomino-Antolín, A.; Rubio-Navarro, A.; Vázquez-Carballo, C.; Herencia, C.; Martín-Sanchez, D.; Farré-Alins, V.; Egea, J.; Cannata, P.; et al. Curcumin reduces renal damage associated with rhabdomyolysis by decreasing ferroptosis-mediated cell death. *FASEB J.* **2019**, *33*, 8961–8975. [CrossRef]
36. Manji, G.A.; Wang, L.; Geddes, B.J.; Brown, M.; Merriam, S.; Al-Garawi, A.; Mak, S.; Lora, J.M.; Briskin, M.; Jurman, M.; et al. PYPAF1, a PYRIN-containing Apaf1-like protein that assembles with ASC and regulates activation of NF-kappa B. *J. Biol. Chem.* **2002**, *277*, 11570–11575. [CrossRef]
37. Thomas, P.G.; Dash, P.; Aldridge, J.R.; Ellebedy, A.H.; Reynolds, C.; Funk, A.J.; Martin, W.J.; Lamkanfi, M.; Webby, R.J.; Boyd, K.L.; et al. The intracellular sensor NLRP3 mediates key innate and healing responses to influenza A virus via the regulation of caspase-1. *Immunity* **2009**, *30*, 566–575. [CrossRef]
38. Allen, I.C.; Scull, M.A.; Moore, C.B.; Holl, E.K.; McElvania-TeKippe, E.; Taxman, D.J.; Guthrie, E.H.; Pickles, R.J.; Ting, J.P.-Y. The NLRP3 inflammasome mediates in vivo innate immunity to influenza A virus through recognition of viral RNA. *Immunity* **2009**, *30*, 556–565. [CrossRef]
39. Gross, O.; Poeck, H.; Bscheider, M.; Dostert, C.; Hannesschläger, N.; Endres, S.; Hartmann, G.; Tardivel, A.; Schweighoffer, E.; Tybulewicz, V.; et al. Syk kinase signalling couples to the Nlrp3 inflammasome for anti-fungal host defence. *Nature* **2009**, *459*, 433–436. [CrossRef]

40. Kanneganti, T.-D.; Body-Malapel, M.; Amer, A.; Park, J.-H.; Whitfield, J.; Franchi, L.; Taraporewala, Z.F.; Miller, D.; Patton, J.T.; Inohara, N.; et al. Critical role for Cryopyrin/Nalp3 in activation of caspase-1 in response to viral infection and double-stranded RNA. *J. Biol. Chem.* **2006**, *281*, 36560–36568. [CrossRef]
41. Bauernfeind, F.G.; Horvath, G.; Stutz, A.; Alnemri, E.S.; MacDonald, K.; Speert, D.; Fernandes-Alnemri, T.; Wu, J.; Monks, B.G.; Fitzgerald, K.A.; et al. Cutting edge: NF-kappaB activating pattern recognition and cytokine receptors license NLRP3 inflammasome activation by regulating NLRP3 expression. *J. Immunol.* **2009**, *183*, 787–791. [CrossRef]
42. Dostert, C.; Pétrilli, V.; Van Bruggen, R.; Steele, C.; Mossman, B.T.; Tschopp, J. Innate immune activation through Nalp3 inflammasome sensing of asbestos and silica. *Science* **2008**, *320*, 674–677. [CrossRef]
43. Cruz, C.M.; Rinna, A.; Forman, H.J.; Ventura, A.L.; Persechini, P.M.; Ojcius, D.M. ATP activates a reactive oxygen species-dependent oxidative stress response and secretion of proinflammatory cytokines in macrophages. *J. Biol. Chem.* **2007**, *282*, 2871–2879. [CrossRef]
44. Ma, M.W.; Wang, J.; Dhandapani, K.M.; Brann, D.W. NADPH Oxidase 2 Regulates NLRP3 Inflammasome Activation in the Brain after Traumatic Brain Injury. *Oxid. Med. Cell. Longev.* **2017**, *2017*, 6057609. [CrossRef]
45. Bao, L.; Huang, Y.; Gu, F.; Liu, W.; Guo, Y.; Chen, H.; Wang, K.; Wu, Z.; Li, J. Zearalenone induces liver injury in mice through ferroptosis pathway. *Sci. Total Environ.* **2024**, *952*, 175875. [CrossRef]

animals

MDPI

Article

Effects of Low-Protein Diet Without Soybean Meal on Growth Performance, Nutrient Digestibility, Plasma Free Amino Acids, and Meat Quality of Finishing Pigs

Bobo Deng [1,2], Litong Wang [2], Xiaomei Jiang [2], Tianyong Zhang [2], Mingfei Zhu [2], Guoshui Wang [2,*], Yizhen Wang [1] and Yuanzhi Cheng [1,*]

1 College of Animal Sciences, Zhejiang University, Hangzhou 310058, China; 18168960899@163.com (B.D.); yzwang321@zju.edu.cn (Y.W.)

2 Zhejiang Tianpeng Group Co., Ltd., Quzhou 324014, China; 18367079999@163.com (L.W.); m18367070887@163.com (X.J.); ztyd2009@163.com (T.Z.); 13260642771@163.com (M.Z.)

* Correspondence: 18357089999@163.com (G.W.); yzcheng@zju.edu.cn (Y.C.)

Simple Summary: This research investigated the impact of a low—protein diet without soybean meal on finishing pigs. The results showed that this diet didn't adversely affect the pigs' growth performance. It enhanced the digestibility of dry matter, crude protein, and calcium. Although there were some changes in plasma free amino acids, the meat quality of the pigs wasn't negatively influenced. Moreover, this diet led to a reduction in ammonia—nitrogen and hydrogen sulfide levels in the pigs' urine. In essence, feeding finishing pigs a low—protein diet without soybean meal can maintain their growth and meat quality while possibly cutting feed costs and lessening environmental impact.

Abstract: This study investigated the effects of a low-protein diet without soybean meal on the growth performance, nutrient digestibility, plasma-free amino acids, and meat quality of finishing pigs. A total of 81 barrows and 81 gilts ([Yorkshire × Landrace] × Duroc) of 150 days old with similar body weights (103.64 ± 3.65 kg) were selected and randomly allocated into three groups with three replicates in each group and 18 finishing pigs in each pen (4.5 m × 8.5 m). Three dietary formulations were established in this experiment: (1) 14% crude protein, and 14% soybean meal for the control group diet (CON); (2) 13% crude protein, and 7% soybean meal for half soybean meal group diet (HSB); (3) 12% crude protein, and 0% soybean meal for soybean meal free group diet (SBF). Three groups of diets were equal to standard ileal digestible amino acids and swine net energy. Results showed that (1) there was no significant difference in final body weight, average daily feed intake, and average daily gain in CON, HSB, and SBF groups ($p > 0.05$). However, the feed conversion rate was higher in the SBF but the differences were not significant ($p = 0.08$). (2) Compared with the CON group, the apparent digestibility of dry matter, crude protein, and calcium in the SBF group was significantly increased by 6.42%, 4.78%, and 9.82%, respectively ($p < 0.05$). However, ether extract and phosphorus did not differ significantly between the three groups ($p > 0.05$). All groups had no significant differences in serum biochemical indexes ($p > 0.05$). (3) Compared with the CON group, the plasma-free contents of glycine and glutamate in the HSB and SBF groups were significantly decreased by 42.90% and 37.06%, respectively ($p < 0.05$). Compared with the CON group, the contents of plasma-free histidine and valine in the SBF group were significantly decreased by 16.72% and 16.21%, respectively ($p < 0.05$), but there was no significant difference compared with the HSB group ($p > 0.05$). Compared with the CON group, plasma-free isoleucine content in the HSB group was significantly decreased by 50.39% ($p < 0.05$), but there was no significant difference compared with the SBF group ($p > 0.05$). (4) Backfat thickness in the SBF group was significantly increased by 3.53% compared with that in the CON group ($p < 0.05$), but there was no significant difference compared with the HSB group ($p > 0.05$). Compared with the CON group, lean meat percentage in the HSB and SBF groups showed a tendency to increase and decrease, respectively ($p = 0.09$). (5) The urine ammonia-nitrogen content of finishing pigs in the SBF group was significantly lower than that of the CON ($p < 0.05$), at a rate of 28.80%. H_2S content in the urine of the SBF group tended to decrease ($p = 0.06$). This study demonstrates that a low-protein diet without soybean meal can be applied as a dietary strategy for finishing pigs with no adverse influence on growth performance.

Keywords: low-protein diet; soybean meal; growth performance; nitrogen utilization; meat quality; finishing pigs; gut health

1. Introduction

With the development of animal husbandry, the shortage of protein feed resources has become increasingly prominent, especially that of soybean meal as the main protein source, since its price fluctuation and supply stability have a great impact on the livestock industry [1,2]. Therefore, it is of great significance to study the application of a low-protein diet without soybean meal in pigs to reduce feed cost, improve feed utilization rate, and ensure the sustainable development of animal husbandry. The trial period was chosen to focus on the later stage of finishing pigs' growth, which is an important period for assessing the impact of dietary changes on growth performance and meat quality. Soybean meal is one of the main protein raw materials in the pig diet, but with the rising price in recent years, soybean meal reduction alternatives have been paid more and more attention in China [3,4]. In pig production, low crude protein (CP) formulation with supplemental crystalline amino acid (CAA) has been widely implemented as a strategy to reduce the feed cost, and more recently to improve intestinal health mainly by decreasing the protein fermentation by intestinal microbiota [5–7]. This practice requires a higher energy-to-protein ratio in the diet to support rapid growth and fat deposition [8]. Therefore, a soybean meal-free diet with optimized protein levels can be particularly beneficial in this context. Reducing the dietary CP by 2% to 4% compared to the NRC (2012) [9] recommendations and supplementing with CAA has been demonstrated to increase nitrogen utilization, reduce feed costs and nitrogen excretion, and promote gut health without impairing the growth performance of pigs [10–12]. Robert et al. showed that reducing dietary CP content does not negatively impact the digestion of nutrients in the foregut of pigs [13]. Wu et al. reduced the dietary CP level of pigs by 1–3% and found that the growth performance of the pigs was not affected, and the NH_3 content as well as nitrogen excretion in the fattening pig house were reduced regardless of the growth stage or fattening stage [14]. Li et al. showed that a 3% reduction in dietary CP level could increase the intramuscular fat content of the *longissimus dorsi* muscle of fattening pigs, reduce shear force, and help to improve meat quality [15]. However, significantly cutting down on dietary CP might negatively impact growth performance, as the synthesis of non-essential amino acids (AAs) and the availability of bioactive compounds from protein supplements could be restricted, even if the AA requirements are satisfied [16,17]. What's more, using a low CP formula might lead to an increase in the net energy content of feeds, thereby causing excessive fat deposition [18]. On the other hand, adding more functional AAs to the low CP formula can further improve intestinal health and glucose metabolism, which in turn enhances nitrogen utilization and growth performance [19]. Although low-protein diets have been studied extensively in pigs, there are few studies based on low-protein diets without soybean meal. It is worth noting that in southern China, the selling price of 150 kg fattened pigs is significantly higher than that of 120 kg standard pigs, and breeding enterprises and family farms have a habit of raising large, fat pigs. According to the nutritional requirements of pigs, it is the most likely stage to use formula without soybean meal. Thus, a low-protein strategy can be realized.

There were few studies examining the application of a low-protein diet without soybean meal in the last stage of the fattening period; however, its effect on finishing pigs is still unknown. Hence, this work was conducted to evaluate the effects of a low-protein diet without soybean meal on growth performance, nutrient digestibility, plasma-free amino acids, and meat quality of finishing pigs. Together, these data could represent a useful basis for applications of a low-protein diet without soybean meal on finishing pigs to lay a good foundation for increasing efficiency and reducing cost.

2. Materials and Methods

A total of 81 barrows and 81 gilts ([Yorkshire × Landrace] × Duroc) of 150 days old with similar body weights (103.64 ± 3.65 kg) were selected and randomly allocated into three groups with 3 replicates in each group and 18 finishing pigs in each pen (4.5 m × 8.5 m). All pigs were housed in an environmentally controlled building with a half-slotted concrete floor facility during finishing periods. The experimental pigs were fed twice at 8:00 a.m. and 5:00 p.m. each day, and the epidemic prevention and disinfection were carried out following the management regulations of the pig farm. The temperature of the pig pens was controlled between 15 and 20 °C and the relative humidity was maintained between 50 and 65% throughout the experiment. Daily feed allocations to each pen were adjusted according to the body weight and the weight of the residual was recorded every day before new feeding was delivered. Fecal samples were collected three times daily at 8 h intervals. After collection, 10% H_2SO_4 was added to fix nitrogen, ammonium sulfate was used as marker, and the samples were stored at −20 °C for 24 h before processing.

The experiment was conducted at the Pig Experiment Research Farm, Zhejiang Tianpeng Group Co., Ltd. (Quzhou, China), from October to December 2023. Following the NRC guidelines from 1998 and 2012, three dietary formulations were established in this experiment (Table 1): (1) 14% crude protein, and 14% soybean meal for the control group diet (CON); (2) 13% crude protein, and 7% soybean meal for half soybean meal group diet (HSB); (3) 12% crude protein, and 0% soybean meal for soybean meal free group diet (SBF). The three dietary groups were equal to standard ileal digestible amino acids and swine net energy. During the whole experiment period, each pen had feeders and nipple drinkers to provide ad libitum access. This experiment consisted of a 7-day adaptation and a 41-day fattening period for sample collection.

Table 1. Ingredients and nutritional contents of the experimental diets.

Items (%)	Treatment [1]		
	CON	HSB	SBF
Ground corn	71.77	45.37	37.03
Wheat	-	25.00	30.00
Soybean meal	14.00	7.00	-
Wheat bran	12.00	10.00	10.00
Rice bran	-	5.00	5.00
Wheat germ meal	-	5.00	15.00
Soybean oil	0.50	0.50	0.50
Salt	0.40	0.40	0.40
Calcium carbonate	0.10	0.10	0.10
$CaHPO_4$	0.50	0.50	0.50

Table 1. *Cont.*

Items (%)	Treatment (1)		
	CON	HSB	SBF
Lysine	0.46	0.52	0.63
Methionine	0.03	0.04	0.05
Threonine	0.10	0.12	0.15
Tryptophan	-	0.01	0.12
Valine	-	-	0.02
Premix (2)	0.14	0.44	0.50
Total	100.00	100.00	100.00
Nutrient Levels (3)			
DM (%)	87.15	87.15	87.15
SNE (Kcal/kg)	2422.00	2418.00	2420.00
CP (%DM)	14.00	13.00	12.00
EE (%DM)	4.12	4.13	4.13
Ash (%DM)	3.50	3.51	3.53
Calcium (%DM)	0.58	0.58	0.58
Phosphorus (%DM)	0.24	0.24	0.24
Swine SID Lys (%)	0.90	0.90	0.90
Swine SID M + C (%)	0.45	0.45	0.45
Swine SID Met (%)	0.24	0.24	0.24
Swine SID Thr (%)	0.50	0.50	0.50
Swine SID Trp (%)	0.13	0.13	0.13
Swine SID Val (%)	0.53	0.53	0.53

Abbreviations: DM = dry matter; SNE = swine net energy; CP = crude protein; EE = ether extract; SID = standardized ileal digestibility; Lys = lysine; Met = methionine; C = cystine; Thr = threonine; Trp = tryptophan; and Val = valine. (1) Treatment: CON = 14% crude protein, and 14% soybean meal; HSB = 13% crude protein, and 7% soybean meal; SBF = 12% crude protein, and 0% soybean meal. (2) Premix was formulated to provide (per kilogram of the dietary DM) 31.5 mg of Zn as $ZnSO_4{\cdot}7H_2O$; 40.1 mg of Mn as $MnSO_4{\cdot}H_2O$; 90.0 mg of Fe as $FeSO_4{\cdot}7H_2O$; 12.0 mg of Cu as $CuSO_4{\cdot}5H_2O$; 0.5 mg of I as KI; 0.3 mg of Co as $CoCl_2{\cdot}6H_2O$; 0.25 mg of Se as Na_2SeO_3; 6800 IU of vitamin A; 4500 IU of vitamin D_3; 200 IU of vitamin E; 9 mg of vitamin K_3, 5.88 mg of vitamin B_1, 4.5 mg of vitamin B_6; and 40 μg of vitamin B_{12}. (3) Swine net energy and SID amino acid were calculated value, while the others were measured values.

2.1. Sample Collection and Determination

The daily feed provided and remnants were collected and weighed to determine the ADFI. All 162 finishing pigs were weighed at the last day of the experiment to calculate final weight, average daily gain, and feed conversion rate. A week-long digestion and metabolism test was carried out at the conclusion of the formal period, with the first two days serving as the adaptation phase and the subsequent five days designated for sampling. Fecal samples were gathered three times daily, with an eight-hour interval between each collection. Subsequent to the collection, 10% sulfuric acid (H_2SO_4) was added to facilitate nitrogen fixation. The five-day fecal samples (200 g) were blended and stored in a refrigerator at 20 °C pending the assessment of the apparent digestibility of nutrients. Both feed and feces were subjected to drying and grinding using a pulverizer to prepare for the subsequent determination of dry DM, CP, EE, Ca, and P content. The detection of DM was performed in accordance with AOAC method 930.15; the measurement of EE was conducted following AOAC method 920.85; and the testing of CP was executed using the Kjeldahl method with an azotometer (Scino KT260, FOSS, Hillerod, Denmark).

At the conclusion of the experiment, six pigs with average body weights were chosen from each group for slaughter. Approximately 5 mL of blood was collected from the jugular vein of each pig into vacuum plasma tubes three hours after morning feeding [20]. One of the blood samples was then centrifuged at 3000× *g* for 10 min at 4 °C to obtain plasma, which was subsequently frozen at −20 °C for further analysis. Serum samples were analyzed for the levels of alanine aminotransferase (ALT), aspartate aminotransferase (AST), alkaline phosphatase (ALP), transglutaminase (GGT), total bilirubin (TBIL), total protein (TP), and so on, following the methods described in previous studies [21]. Another blood sample was used to measure the plasma-free amino acid content.

The weight of the carcass was recorded to calculate the carcass yield. Measurements of backfat thickness at the first rib, last rib, and last lumbar vertebra were taken to determine the average backfat depth. The loin eye area (LEA) was measured at the last lumbar vertebra [22]. Samples of the *longissimus dorsi* (LD) muscle from the 10th ribs were taken to measure the pH value, meat color, shear force, drip loss, cooking loss, moisture loss, and intramuscular fat (IMF) [23,24]. The ultimate pH was measured 24 h after slaughter using an Orion 9106 penetrating probe, which had been calibrated with two buffers (pH 7.00 and 4.01). Cooking loss was calculated 24 h after slaughter by comparing the weight of the cooked sample (B) to the weight of the raw sample (A) using the formula: cooking loss (g/100 g) = $[(A - B)/A] \times 100$. Shear force, expressed as g/100 g of liquid expelled, was measured using the filter paper press method as described by the American Meat Science Association. Meat color was assessed at the same time and location as the pH measurement. The lightness (L*), redness (a*), and yellowness (b*) values were measured three times on the LD muscle for each pig using a Chroma meter (CR–400, Tokyo, Japan), and the average value was used as the result [25]. Shear force was measured using a texture analyzer (TA.XT Plus, Stable Micro Systems, Godalming, UK). The IMF was measured according to the national standard method (GB 5009.6–2016, China) [26]. Urine samples were collected from each pig at 3 h post-feeding. Ammonia-nitrogen levels were measured using the phenol–sodium hypochlorite colorimetric method, while hydrogen sulfide levels were determined using gas chromatography [27].

2.2. Statistical Analysis

The results are shown as the averages along with their standard errors of the mean (SEM). Statistical significance was assessed using one-way ANOVA followed by Tukey's post–hoc tests for multiple comparisons. Before conducting ANOVA, data were tested for normality using the Shapiro–Wilk test and homogeneity of variance using Levene's test. No pigs were eliminated or died during the experiment; hence, all data were included in the analysis. All analyses were performed using the SPSS Statistics software, version 20.0 (IBM Corp., Armonk, NY, USA). In this study, statistical significance was set at $p \leq 0.05$, with highly significant results at $p < 0.01$, and trends were noted when $0.05 < p \leq 0.1$.

3. Results

3.1. Growth Performance

As shown in Table 2, the results showed no significant difference in FBW among all groups ($p > 0.05$). There was also no significant difference in ADFI and ADG among the CON, HSB, and SBF groups ($p > 0.05$). Although the corn content in the SBF group was significantly lower than in other groups, the pigs maintained similar growth performance, suggesting that the optimized protein and energy balance in the diet compensated for the reduced corn levels.

3.2. Nutrient Digestion and Serum Biochemical Parameters

As shown in Table 3, the results showed that compared with the CON group, the apparent digestibility of DM, CP, and Ca in the SBF group was significantly increased by 6.42%, 4.78%, and 9.82%, respectively ($p < 0.05$). However, EE and P did not differ significantly among the three groups ($p > 0.05$). As presented in Table 4, the results showed that there were no significant differences in different serum biochemical indexes among all groups ($p > 0.05$).

Table 2. Influence of low-protein diet without soybean meal on growth performance of finishing pigs.

Items	Treatment			SEM	*p*-Value
	CON (*n* = 3)	**HSB** (*n* = 3)	**SBF** (*n* = 3)		
IBW (kg)	103.28	104.30	103.35	5.03	0.35
FBW (kg)	151.26	153.43	151.48	8.46	0.42
ADFI (g/d)	4209.57	4230.20	4214.09	35.15	0.61
ADG (g/d)	1170.28	1198.25	1173.89	13.10	0.14
FCR	3.60	3.53	3.59	0.02	0.08

Abbreviations: IBW = initial body weight; FBW = final body weight; ADFI = average daily feed intake; ADG = average daily gain; FCR = feed conversion rate.

Table 3. Influence of low-protein diet without soybean meal on nutrient digestion of finishing pigs.

Items	Treatment			SEM	*p*-Value
	CON (*n* = 6)	**HSB** (*n* = 6)	**SBF** (*n* = 6)		
DM (%)	80.10^{b}	82.28^{ab}	86.25^{a}	2.13	<0.05
CP (%)	79.50^{b}	81.10^{ab}	83.45^{a}	1.05	<0.05
EE (%)	75.65	74.60	75.31	3.40	0.38
Ca (%)	55.20^{b}	56.13^{b}	60.62^{a}	2.25	<0.05
P (%)	47.24	51.50	50.68	3.62	0.72

Abbreviations: DM = dry matter; CP = crude protein; EE = ether extract; Ca = calcium; and P = phosphorus. ^{a,b} Means within a row with different superscripts differ ($p < 0.05$). The same as in the following table.

Table 4. Influence of low-protein diet without soybean meal on serum biochemical parameters of finishing pigs.

Items	Treatment			SEM	*p*-Value
	CON (*n* = 6)	**HSB** (*n* = 6)	**SBF** (*n* = 6)		
ALT (U/L)	50.83	52.03	49.96	6.02	0.25
AST (U/L)	35.13	35.30	32.30	3.79	0.42
ALP (U/L)	70.67	71.60	72.33	5.50	0.38
GGT (U/L)	40.00	40.33	42.67	3.26	0.16
TBIL (μmol/L)	7.70	5.23	6.93	0.65	0.22
TP (g/L)	74.73	68.96	75.96	5.32	0.57
ALB (g/L)	41.37	41.26	43.16	3.20	0.32
GLO (g/L)	33.36	27.70	32.80	5.58	0.46
A/G	1.26	1.50	1.33	0.35	0.78
LDH (U/L)	601.00	610.67	607.67	25.65	0.35
CK (U/L)	1426.33	1590.00	1500.67	100.20	0.92
BUN (mmol/L)	4.91	4.54	5.76	0.80	0.20
CREA(μmol/L)	149.20	140.50	146.30	10.25	0.33
GLU (mmol/L)	5.93	6.03	6.23	0.50	0.53
CHOL (mmol/L)	2.31	1.99	2.35	0.04	0.12
TG (mmol/L)	0.43	0.35	0.45	0.08	0.16
PAMY (U/L)	2236.33	2424.00	2380.33	158.35	0.87
LIP (U/L)	17.76	14.70	16.43	3.20	0.28

Abbreviations: ALT = glutamate pyruvate transaminase; AST = glutamic oxalacetic transaminase; ALP = alkaline phosphatase; GGT = transglutaminase; TBIL = total bilirubin; TP = total protein; ALB = albumin; GLO = globulin; A/G = albumin to globulin; LDH = lactic dehydrogenase; CK = creatine kinase; BUN = urea nitrogen; CREA = creatinine; GLU = glucose; CHOL = total cholesterol; TG = triglyceride; PAMY = amylopsin; and LIP = lipase.

3.3. Plasma Free Amino Acid Composition

In this study, in order to explore the effects of low protein without soybean meal diet on plasma-free amino acids of 90–150 kg finishing pigs, the changes in 22 free amino acids in the plasma of CON, HSB, and SBF groups were compared (Table 5). Results demonstrated that compared with the CON group, the contents of glycine and glutamate in the HSB and SBF groups were significantly decreased by 42.90% and 37.06%, respectively ($p < 0.05$). Compared with the CON group, the contents of histidine and valine in the SBF group were significantly decreased by 16.72% and 16.21%, respectively ($p < 0.05$), but there was no significant difference compared with the HSB group ($p > 0.05$). Compared with the CON group, isoleucine content in the HSB group was significantly decreased by 50.39% ($p < 0.05$), but there was no significant difference compared with the SBF group ($p > 0.05$). The contents of other amino acids were not significantly different among the three groups ($p > 0.05$).

Table 5. Influence of low-protein diet without soybean meal on plasma-free amino acid composition in finishing pigs.

Items	Treatment			SEM	p-Value
	CON (n = 6)	HSB (n = 6)	SBF (n = 6)		
Glycine (μg/mL)	219.10 ^{a}	125.10 ^{b}	137.91 ^{b}	10.03	<0.05
Serine (μg/mL)	253.04	248.56	246.21	6.46	0.12
Methionine (μg/mL)	72.33	63.64	66.59	5.15	0.51
Proline (μg/mL)	28.36	25.72	24.37	3.20	0.24
Leucine (μg/mL)	33.55	30.78	31.64	2.12	0.78
Creatine (μg/mL)	43.71	33.76	42.53	6.28	0.37
Glutamate (μg/mL)	229.20 ^{a}	165.13 ^{b}	173.69 ^{b}	10.20	<0.05
Phenylalanine (μg/mL)	18.67	12.89	16.64	3.50	0.13
Lysine (μg/mL)	56.96	52.71	54.34	3.78	0.62
Argine (μg/mL)	25.39	21.86	20.89	4.10	0.39
Tryptophan (μg/mL)	15.18	12.85	15.76	2.30	0.17
Tyrosine (μg/mL)	13.01	11.49	12.02	1.80	0.59
Histidine (μg/mL)	47.67 ^{a}	43.06 ^{ab}	39.70 ^{b}	0.90	<0.05
Valine (μg/mL)	142.68 ^{a}	138.52 ^{ab}	119.55 ^{b}	8.15	<0.05
Ornithine (μg/mL)	57.42	50.09	52.12	6.05	0.17
Alanine (μg/mL)	40.59	33.83	39.92	5.26	0.82
Taurine (μg/mL)	31.83	31.99	35.33	3.80	0.28
Isoleucine (μg/mL)	23.10 ^{a}	11.46 ^{b}	17.23 ^{ab}	4.30	<0.05
Aspartic (μg/mL)	10.52	8.61	9.03	1.28	0.47
Threonine (μg/mL)	42.64	39.46	33.93	5.90	0.62
Glutamine (μg/mL)	58.21	54.72	56.06	3.02	0.15
Asparagine(μg/mL)	13.89	14.62	16.27	2.40	0.17

^{a,b} Means within a row with different superscripts differ ($p < 0.05$). The same as in the following table.

3.4. Meat Quality

As is shown in Table 6, backfat thickness in the SBF group was significantly increased by 3.53% compared with that in the CON group ($p < 0.05$), but there was no significant difference compared with the HSB group ($p > 0.05$). Compared with the CON group, lean meat percentage in the HSB and SBF groups tended to increase and decrease, respectively ($p = 0.09$). The indexes of slaughter rate and loin eye area showed no significant differences among the three groups ($p > 0.05$). There were no significant differences in pH value, meat color (a*, b*, L*), shear force, drip loss, cooking loss, milling loss, or intramuscular fat content indexes of *longissimus dorsi* muscle ($p > 0.05$).

Table 6. Influence of low-protein diet without soybean meal on carcass traits and meat quality of finishing pigs.

Items	Treatment			SEM	p-Value
	CON (n = 6)	HSB (n = 6)	SBF (n = 6)		
Carcass traits					
YP (%)	72.83	70.25	71.95	1.32	0.56
CL (cm)	91.05	90.50	92.89	2.58	0.42
BFT (cm)	2.83 ^{b}	2.87 ^{ab}	2.93 ^{a}	0.05	<0.05
LEA (cm^2)	32.58	33.06	32.73	2.80	0.17
LMR (%)	60.20	61.52	57.90	2.35	0.09
Meat quality					
pH_{24h}	5.72	5.68	5.70	0.15	0.35
a*	16.30	16.25	16.81	1.80	0.57
b*	4.23	4.50	4.46	0.45	0.81
L*	56.10	57.20	56.52	3.55	0.70
SF (N)	37.00	36.50	37.87	4.90	0.32
DL (%)	4.65	4.70	4.62	0.89	0.61
CL (%)	15.50	14.92	15.10	0.80	0.20
ML (%)	37.10	37.45	38.06	10.25	0.38
IMF (%)	1.80	1.84	1.95	0.50	0.53

Abbreviations: YP = yield percentage; CL = carcass length; BFT = back fat thickness; LEA = loin eye area; LMR = lean meat rate; SF = shear force; DL = drip loss; CL = cooking loss; ML = milling loss; and IMF = intramuscular fat. ^{a,b} Means within a row with different superscripts differ ($p < 0.05$). The same as in the following table.

3.5. The Content of H_2S and Ammonia-Nitrogen in Urine

As is shown in Table 7, by adding no soybean meal to the diet of the SBF group, the urine ammonia-nitrogen content of 90–150 kg finishing pigs in the SBF group was significantly lower than that of the CON at the end of the experiment ($p < 0.05$), decreased by 28.80%. H_2S content in the urine of the SBF group tended to decrease but there was no significant difference between the two groups ($p = 0.06$).

Table 7. Influence of low-protein diet without soybean meal on the content of H_2S and ammonia-nitrogen in urine of finishing pigs.

Items	Treatment			SEM	p-Value
	CON (n = 6)	HSB (n = 6)	SBF (n = 6)		
NH_3-N (mg/dL)	1.25 ^{a}	1.14 ^{ab}	0.89 ^{b}	0.13	<0.05
H_2S (nmol/mL)	5.25	5.16	4.12	0.75	0.06

^{a,b} Means within a row with different superscripts differ ($p < 0.05$). The same as in the following table.

4. Discussion

4.1. Growth Performance, Nutrient Digestion, and Serum Biochemical Parameters

Growth performance is the most direct metric in livestock production. In this study, it was observed that feeding a low-protein diet, with crude protein reduced from 14% to 12%, had no negative impact on the growth performance of finishing pigs. Han et al. reported that in the growing stage, pigs fed a low-protein diet according to the NRC (2012) recommendations showed a decrease in average daily gain (ADG) [28]. Jiang et al. showed that supplementing glycine to a low-protein diet could improve meat quality without affecting animal growth [29]. Duan et al. found that when the amino acid requirements of growing–finishing pigs were met, reducing the dietary protein level by

2–4% did not affect the growth performance of the pigs [30]. Wang et al. demonstrated that pigs fed a low-protein diet might have longer villi in the small intestine during the finishing phase compared to those fed higher protein diets, which could lead to improved ADG [31,32]. Serum biochemical indicators can reflect changes in physiological functions and health status of the animal, as well as provide information on nutritional metabolism and organ function changes. The levels of serum albumin and total protein are also indicators of the animal's health status [33]. In this study, we found that the low-protein diet without soybean meal had no negative effect on the blood biochemical parameters and thus the health status of finishing pigs. It should be noted that in this study, there was a significant substitution of maize with wheat in the diets. This change may have an impact on the results, as wheat and maize have different nutritional profiles and may affect the digestibility and utilization of nutrients. Future studies should consider this factor and investigate the effects of different grain sources on the performance of finishing pigs.

4.2. Free Amino Acid Profile in Plasma

The level of protein in the diet directly affects the metabolism of protein in the pig body, and reducing the level of protein in the diet can improve the metabolism of protein [33]. Amino acids are the basic components of protein and an important part of life activities. As an important metabolite of animal metabolism, plasma amino acids reflect the status of animal protein metabolism and energy metabolism [34]. Bergen et al. demonstrated that the content of free amino acids in plasma could reflect the metabolism of amino acids in animals. There was no significant difference in free amino acids in plasma, indicating that the ratio of various amino acids in the diet was balanced [35]. Ye et al. showed that after the reduction in dietary CP level, the plasma-free amino acids of finishing pigs were significantly reduced compared to that of the control group, except for histidine, valine, and isoleucine, which had similar content among groups [36]. In this study, the content of histidine and valine in plasma was decreased by the low-protein diet without soybean meal. It is indicated that the ratio of amino acids in the diet was basically balanced, and the metabolism and utilization of amino acids were not affected by the decrease of CP from 14 to 12%. The significant decrease in glycine and glutamate levels in the SBF group may reflect altered metabolic pathways in response to the low-protein diet. These changes could potentially affect energy metabolism and muscle development, although no adverse effects on growth performance were observed in this study. Especially synthetic AAs, such as L-lysine-HCl, DL-methionine, L-threonine, and tryptophan, were supplemented to increase nitrogen utilization and to reduce negative effects deriving from a reduction in CP. We hypothesized that when the above four amino acids are in sufficient levels in the diet, the reduction in crude protein would cause the finishing pigs to be more sensitive to the deficiency of histidine and valine.

4.3. Meat Quality and the Content of H_2S and Ammonia-Nitrogen in Urine

There have been many reports on the influence of dietary protein levels on meat quality. The main indexes to evaluate meat quality are meat color, marbling, intramuscular fat, and pH value [37,38]. Li et al. found that lower levels of protein in the diet decreased shear force (improved tenderness), probably through an increase in total fat content and modification in fatty acid composition in the muscles of growing pigs [15]. Zhu et al. showed that a low-protein amino acid balanced diet could improve the tenderness of *longissimus dorsi* muscle, increase intramuscular fat content, and change the fatty acid composition of fattening pigs [39]. In our study, we found that backfat thickness in the SBF group was significantly increased by 3.53% compared with that in the CON group, but there were no significant differences in pH value, meat color (a*, b*, L*), shear force, drip loss, cooking loss, milling loss, and intramuscular fat content indexes of *longissimus dorsi* muscle when the proportion of soybean meal decreased from 14% to 0, indicating that low-protein diet without soybean meal did not have adverse effects on the meat quality of 90–150 kg finishing pigs. The levels of NH_3-N and H_2S in the urine not only affected the health of the livestock and humans but also reflected the nitrogen utilization capacity of animals [40,41]. In this study, NH_3-N in the urine of finishing pigs fed with a low-protein diet without soybean meal was significantly decreased and the content of H_2S in the excreta tended to be reduced.

5. Conclusions

In summary, under the conditions of this experiment, offering a low-protein diet without soybean meal to finishing pigs had no significant influence on the growth performance, but elevated the apparent digestibility of dry matter, crude protein, and calcium. In addition, there was no adverse effect on carcass traits, meat quality, and blood biochemical parameters on finishing pigs. Moreover, a low-protein diet without soybean meal reduced the emission of both ammoniacal nitrogen and sulfuretted hydrogen in urine. Therefore, a low-protein diet without soybean meal can be applied as a dietary strategy for finishing pigs, which can reduce the amount of soybean meal purchased annually for enterprises.

Author Contributions: B.D., Y.W. and Y.C. designed the whole experiment and verified the validity of experiment and checked the results. X.J., T.Z. and M.Z. performed the experiment, including chemical analysis, and statistical analysis. B.D. worked on the manuscript. L.W. and G.W. participated in the experiment design and gave valuable advice. Y.C. revised the manuscript and checked the data. All authors have read and agreed to the published version of the manuscript.

Funding: The authors thank the specialized research fundings from the Fundamental Research Funds for the Central Universities (226–2024-00240), National Key Research and Development Plan (2023YFD1301303), Zhenong Young Talents Program, Taishan Industrial Leading Talents Project, the National Center of Technology Innovation for Pigs, the China Agriculture Research System of MOF, and MARA (CARS-35).

Institutional Review Board Statement: This experiment was performed under protocols approved by the Animal Care and Use Committee of Zhejiang University, Hangzhou, China. The authors confirm that they have followed EU standards for the protection of animals used for scientific purposes.

Informed Consent Statement: Not applicable.

Data Availability Statement: The original contributions presented in this study are included in the article. Further inquiries can be directed to the corresponding author(s).

Conflicts of Interest: Authors Bobo Deng, Litong Wang, Xiaomei Jiang, Tianyong Zhang, Mingfei Zhu and Guoshui Wang were employed by the Zhejiang Tianpeng Group Co., Ltd. The remaining authors declare that the research was conducted in the absence of any commercial or financial relationships that could be construed as a potential conflict of interest.

References

1. Li, Y.; Yin, J.; Han, H. Metabolic and proteomic responses to long-term protein restriction in a pig model. *J. Agric. Food Chem.* **2018**, *47*, 12571–12579. [CrossRef] [PubMed]
2. Wang, Y.; Zhou, J.; Cao, N. Dietary crude protein time-dependently modulates the bacterial community and metabolites and changes dietary nutrient efficiency in growing pigs. *Anim. Nutr.* **2023**, *17*, 1–10. [CrossRef] [PubMed]
3. Grabež, V.; Egelandsdal, B.; Kjos, N.P. Replacing soybean meal with rapeseed meal and faba beans in a growing-finishing pig diet: Effect on growth performance, meat quality and metabolite changes. *Meat Sci.* **2020**, *166*, 108134. [CrossRef]
4. Pirgozliev, V.R.; Whiting, I.M.; Mansbridge, S.C. Sunflower and rapeseed meal as alternative feed materials to soybean meal for sustainable egg production, using aged laying hens. *Br. Poult. Sci.* **2023**, *5*, 634–640. [CrossRef]

5. Wang, Y.; Zhou, J.; Qiao, S. Advances in low-protein diets for swine. *J. Anim. Sci. Biotechnol.* **2018**, *9*, 60. [CrossRef]
6. Wang, Y.; Han, S.; Zhou, J. Effects of dietary crude protein level and N-carbamylglutamate supplementation on nutrient digestibility and digestive enzyme activity of jejunum in growing pigs. *J. Anim. Sci.* **2020**, *4*, skaa088. [CrossRef]
7. Nyachoti, C.M.; Omogbenigun, F.O.; Rademacher, M.; Blank, G. Performance responses and indicators of gastrointestinal health in early-weaned pigs fed low-protein amino acid-supplemented diets. *J. Anim. Sci.* **2006**, *1*, 125–134. [CrossRef]
8. Liu, S.; Fan, Z. Effects of dietary protein restriction on colonic microbiota of finishing pigs. *Animals* **2022**, *1*, 9. [CrossRef]
9. National Research Council (NRC). *Nutrient Requirements of Swine*; National Academy Press: Washington, DC, USA, 1998.
10. Niyonsaba, A.; Jin, X.H.; Kim, Y.Y. Effect of reducing dietary crude protein level on growth performance, blood profiles, nutrient digestibility, carcass traits, and odor emissions in growing-finishing pigs. *Anim. Biosci.* **2023**, *10*, 1584–1595. [CrossRef]
11. Fan, P.; Liu, P.; Song, P. Moderate dietary protein restriction alters the composition of gut microbiota and improves ileal barrier function in adult pig model. *Sci. Rep.* **2017**, *7*, 43412. [CrossRef]
12. Kerr, B.J.; Southern, L.L.; Bidner, T.D.; Friesen, K.G.; Easter, R.A. Influence of dietary protein level, amino acid supplementation, and dietary energy levels on growing-finishing pig performance and carcass composition. *J. Anim. Sci.* **2003**, *12*, 3075–3087. [CrossRef]
13. Pieper, R.; Kröger, S.; Richter, J.F.; Wang, J.; Martin, L.; Bindelle, J.; Htoo, J.K.; von Smolinski, D.; Vahjen, W.; Zentek, J.; et al. Fermentable fiber ameliorates fermentable protein-Induced changes in microbial ecology, but not the mucosal response, in the colon of piglets. *J. Nutr.* **2012**, *4*, 661–667. [CrossRef] [PubMed]
14. Wu, D.; Zhao, H.; Chen, S. Effects of low protein diet supplemented with amino acids on growth performance and nitrogen excretion of growing-finishing pigs. *Anim. Husb. Feed. Sci.* **2010**, *5*, 3.
15. Li, Y.H.; Li, F.N.; Duan, Y.H. Low-protein diet improves meat quality of growing and finishing pigs through changing lipid metabolism, fiber characteristics, and free amino acid profile of the muscle. *J. Anim. Sci.* **2018**, *8*, 3221–3232. [CrossRef] [PubMed]
16. Duarte, M.E.; Parnsen, W.; Zhang, S.; Abreu ML, T.; Kim, S.W. Low crude protein formulation with supplemental amino acids for its impacts on intestinal health and growth performance of growing-finishing pigs. *J. Anim. Sci. Biotechnol.* **2024**, *1*, 55. [CrossRef]
17. Kim, S.W.; Chen, H.; Parnsen, W. Regulatory Role of Amino Acids in Pigs Fed on Protein-restricted Diets. *Curr. Protein Pept. Sci.* **2019**, *2*, 132–138. [CrossRef]
18. Pan, S.; Jia, Y.; Yang, X. Amino acid starvation-induced autophagy is involved in reduced subcutaneous fat deposition in weaning piglets derived from sows fed low-protein diet during gestation and lactation: Autophagy is involved in reduced fat deposition in maternal low-protein piglets. *Eur. J. Nutr.* **2018**, *3*, 991–1001.
19. Liu, X.; Pan, S.; Li, X. Maternal low-protein diet affects myostatin signaling and protein synthesis in skeletal muscle of offspring piglets at weaning stage. *Eur. J. Nutr.* **2015**, *6*, 971–979. [CrossRef]
20. Corino, C.; Musella, M.; Mourot, J. Influence of extruded linseed on growth, carcass composition, and meat quality of slaughtered pigs at one hundred ten and one hundred sixty kilograms of liveweight. *J. Anim. Sci.* **2008**, *8*, 1850–1860. [CrossRef]
21. Davidson, S.; Odle, J. Supplementing limited methionine diets with rumen-protected methionine, betaine, and choline in early lactation Holstein cows. *J. Dairy Sci.* **2008**, *91*, 1552–1559. [CrossRef]
22. Yang, Y.; Guo, J.; Kim, J. Effects of growth rate on carcass and meat quality traits and their association with metabolism-related gene expression in finishing pigs. *Anim. Sci. J.* **2012**, *2*, 169–177. [CrossRef] [PubMed]
23. Li, Y.; Tao, X.; Zhao, P. Effects of slaughter weight on carcass characteristics, meat quality, and metabolomics profiling in the *longissimus dorsi* muscle of Tianfu finishing pigs. *Front. Vet. Sci.* **2024**, *11*, 1420634. [CrossRef] [PubMed]
24. Honikel, K.O. Reference methods for the assessment of physical characteristics of meat. *Meat Sci.* **1998**, *49*, 447–457. [CrossRef] [PubMed]
25. Ji, J.; Zhou, L.; Huang, Y. A whole-genome sequence based association study on pork eating quality traits and cooking loss in a specially designed heterogeneous F6 pig population. *Meat Sci.* **2018**, *146*, 160. [CrossRef]
26. *GB 5009.6-2016*; National Food Safety Standard-Determination of Fat in Food. The National Health and Family Planning Commission of the People's Republic of China and State Food and Drug Administration: Beijing, China, 2016.
27. Maiteseidi, T.; Yibureyimu, A. Determination of ammonia-nitrogen concentration in rumen fluid treated with methanol by sodium hypochlorite and phenol spectrophotometry. *Xinjiang Agric. Sci.* **2012**, *3*, 179–184.
28. Han, Y.G.; Lee, G.I.; Do, S.H. The Effect of Reduced Crude Protein on Growth Performance, Nutrient Digestibility, and Meat Quality in Weaning to Finishing Pigs. *Animals* **2023**, *12*, 1938. [CrossRef]
29. Jiang, S.; Quan, W.; Luo, J. Low-protein diets supplemented with glycine improves pig growth performance and meat quality: An untargeted metabolomic analysis. *Front. Vet. Sci.* **2023**, *10*, 1170573. [CrossRef]
30. Duan, J.Q.; Ding, P.; Wu, S.B. Effects of low protein diet on growth performance, carcass quality and plasma amino acids of pigs. *Chin. J. Anim. Sci.* **2020**, *12*, 99–103.
31. Wang, M.; Yang, C.; Wang, Q. The Relationship between Villous Height and Growth Performance, Small Intestinal Mucosal Enzymes Activities and Nutrient Transporters Expression in Weaned Piglets. *J. Anim. Physiol. Anim. Nutr.* **2020**, *104*, 606–615. [CrossRef]

32. Magowan, E.; McCann, M.E.E.; Beattie, V.E. Investigation of Growth Rate Variation between Commercial Pig Herds. *Animal* **2007**, *1*, 1219–1226. [CrossRef]
33. Salazar-Villanea, S.; Hendriks, W.H.; Bruininx, E.M. Protein structural changes during processing of vegetable feed ingredients used in swine diets: Implications for nutritional value. *Nutr. Res. Rev.* **2016**, *1*, 126–141. [CrossRef] [PubMed]
34. Gotti, R.; Esposito, E.; Luise, D. Determination of Free Amino Acids in Milk, Colostrum and Plasma of Swine via Liquid Chromatography with Fluorescence and UV Detection. *Molecules* **2022**, *13*, 4153. [CrossRef] [PubMed]
35. Bergen, W.G. Amino Acids in Beef Cattle Nutrition and Production. *Adv. Exp. Med. Biol.* **2021**, *12*, 29–42.
36. Ye, C.; Zeng, X.; Zhu, J. Dietary N-Carbamylglutamate Supplementation in a Reduced Protein Diet Affects Carcass Traits and the Profile of Muscle Amino Acids and Fatty Acids in Finishing Pigs. *J. Agric. Food Chem.* **2017**, *28*, 5751–5758. [CrossRef]
37. Wang, Q.; Gao, H.; Fu, Y. Comprehensive characterization of the differences in metabolites, lipids, and volatile flavor compounds between Ningxiang and Berkshire pigs using multi-omics techniques. *Food Chem.* **2024**, *457*, 139807. [CrossRef]
38. Fu, Y.; Cao, S.; Yang, L. Flavor formation based on lipid in meat and meat products: A review. *J. Food Biochem.* **2022**, *12*, 14439. [CrossRef]
39. Zhu, Y.P. Effects of Low Protein Amino Acid Balance Diet and Cysteamine Supplementation on Pork Quality. Ph.D. Thesis, Nanjing Agricultural University, Nanjing, China, 2017.
40. Li, H.; Xin, H.; Burns, R.T. Reducing ammonia emissions from laying-hen houses through dietary manipulation. *J. Air Waste Manag. Assoc.* **2012**, *2*, 160–169. [CrossRef]
41. Abd El-Hack, M.E.; Mahgoub, S.A.; Alagawany, M. Improving productive performance and mitigating harmful emissions from laying hen excreta via feeding on graded levels of corn DDGS with or without Bacillus subtilis probiotic. *J. Anim. Physiol. Anim. Nutr.* **2017**, *5*, 904–913. [CrossRef]

animals

MDPI

Article

Tea Polyphenols Mitigate TBBPA-Induced Renal Injury Through Modulation of ROS-PI3K/AKT-NF-κB Signalling in Carp (*Cyprinus carpio*)

Fuxin Han, Ran Xu, Hongru Wang, Xuejiao Gao * and Mengyao Guo *

Department of Clinical Veterinary Medicine, College of Veterinary Medicine, Northeast Agricultural University, Harbin 150030, China; s230601062@neau.edu.cn (F.H.); xr1998_lucky@163.com (R.X.); s230601054@neau.edu.cn (H.W.)

* Correspondence: xuejiaogao@126.com (X.G.); gmy@neau.edu.cn (M.G.)

Simple Summary: As a food-derived bioactive compound, TPs have garnered extensive attention for their physiological properties in research. In this study, we found that TPs lower ROS levels, restore antioxidant enzymes (SOD, CAT), and inhibit the ROS-PI3K/AKT-NF-κB pathway, preventing inflammatory injury. TPs also suppress apoptosis by regulating Caspase-3, Bax, and Bcl-2, and relieve necrosis by downregulating the RIPK3/MLKL signalling pathway. These findings suggest that dietary TPs may serve as a therapeutic agent against TBBPA-induced renal injury in common carp.

Abstract: Tetrabromobisphenol A (TBBPA), a widely utilised brominated flame retardant, demonstrates toxicological effects in aquatic organisms. Tea polyphenols (TPs), natural compounds found in tea leaves, exhibit both antioxidant and anti-inflammatory activities. The kidney is one of the major metabolic organs in common carp and serves as a target organ for toxic substances. This study evaluated the therapeutic potential of TPs in mitigating TBBPA-induced nephrotoxicity in common carp. Common carp were exposed to 0.5 mg/L TBBPA in water and/or fed a diet supplemented with 1 g/kg TPs for 14 days. In vitro, primary renal cells were treated with 60 μM TBBPA and/or 2.5 μg/L TPs for 24 h. Methods included histopathology, TUNEL assay for apoptosis, ROS detection, and molecular analyses. Antioxidant enzymes (SOD, CAT) and inflammatory cytokines (IL-1β, IL-6, TNF-α) were quantified using ELISA kits. Results showed that TBBPA induced oxidative stress, and activated the ROS-PI3K/AKT-NF-κB pathway, thereby resulting in inflammatory responses. TBBPA upregulated apoptosis-related genes (*Caspase-3*, *Bax*, and *Bcl-2*) and induced apoptosis. TBBPA upregulated the expression of RIPK3/MLKL, thereby exacerbating necroptosis. TPs intervention significantly mitigated these effects by reducing ROS, suppressing NF-κB activation, and restoring antioxidant enzyme activities (SOD, CAT). Moreover, TPs attenuated apoptosis and necrosis in the carp kidney, thereby enhancing the survival ability and immunity of common carp.

Keywords: tea polyphenols; aquaculture feed additives; inflammatory injury; oxidative stress

1. Introduction

Common carp (*Cyprinus carpio* L.) ranks as the third most cultivated freshwater fish globally. More than 90% of the global output comes from Asia, and it is one of the freshwater fish commonly consumed by Chinese people [1]. Common carp provides humans with quality animal protein and essential minerals, including iron and zinc. The renal system in carp plays a crucial role in the excretion of metabolic waste. Its function often makes it a target for toxic substances that cause tissue damage [2]. Kidney damage in carp prolongs the breeding cycle and even increases the mortality rate, resulting in greater difficulty in feeding and management [3].

Tetrabromobisphenol A (TBBPA), a widely used brominated flame retardant, is incorporated as an additive in plastics, textiles, and electronic equipment [4,5]. Due to its lipophilicity and thermal stability, TBBPA and its derivatives exhibit significant bioaccumulation and biomagnification in organisms, resulting in developmental, reproductive, and endocrine disruptions, particularly in rodent models and aquatic organisms [6–8]. Environmental monitoring has identified TBBPA at varying concentrations in air, sediments, and marine systems worldwide [9]. Asia accounts for approximately 80% of global TBBPA consumption, with China being the dominant consumer [10,11]. Aquatic sediments serve as a reservoir for TBBPA, and its toxic effects on marine organisms have been widely documented [12]. Yang et al. documented TBBPA concentrations as high as 4.87 μg/L in Chaohu Lake, representing the highest reported aqueous environmental concentration worldwide, particularly in grass carp (*Ctenopharyngodon idella*) and rodent models [13]. TBBPA bioaccumulation was detected in grass carp tissues from Chaohu Lake, with the highest concentrations observed in the kidney (75.2–162.4 ng/g dw), followed by the liver and muscle. TBBPA can activate the ROS/NF-κB pathway, leading to intestinal inflammation and cellular necrosis in common carp [14]. In vitro studies using grass carp hepatocytes have demonstrated that TBBPA induces apoptosis by elevating reactive oxygen species (ROS) levels and impairing ER-mitochondrial communication [15]. Fukuda et al. observed that neonatal rats orally administered 200 mg/kg and 600 mg/kg of TBBPA for 18 days developed multicystic kidney lesions, confirming its nephrotoxicity [16]. TBBPA-induced nephrotoxicity in rodents is primarily characterised by renal tubular abnormalities and cysts, as well as oxidative stress-mediated tissue injury, with juvenile individuals exhibiting higher susceptibility. A high dosage (typically $\geq$250 mg/kg bw) is a critical factor in triggering renal toxicity [16–18]. However, the mechanisms underlying TBBPA-induced renal injury in teleost fish remain to be elucidated.

The PI3K/AKT/NF-κB pathway plays a dual role in teleost kidney physiology, regulating immune responses and tissue repair while also contributing to renal injury and fibrosis when dysregulated. Paraquat exposure leads to apoptosis and programmed necrosis through oxidative stress and the PTEN/PI3K/AKT pathway, thereby causing immune dysfunction in CIK cells [19]. Albicanol antagonises hepatocyte apoptosis induced by Profenofos exposure through a ROS-mediated PTEN/PI3K/AKT pathway [20]. Tea polyphenols (TPs), the most abundant and biologically active natural compounds in Camellia sinensis leaves, exhibit significant physiological properties including antioxidant, anti-inflammatory, and immunomodulatory activities [21]. As food-derived bioactive compounds, TPs have garnered considerable attention in the development of functional feeds. Dietary supplementation with TPs has been shown to mitigate oxidative stress in the gills of common carp, reduce inflammatory responses, and enhance innate immune function [22]. Zhao et al. demonstrated that TPs attenuate ethionine-induced oxidative stress, inflammatory response, and apoptosis in Ctenopharyngodon idella kidney (CIK) cells by suppressing the ROS/MAPK/NF-κB pathway [23]. However, few studies have investigated the therapeutic effects of TPs on renal injury in carp, leaving it unclear if TPs can reduce TBBPA-induced renal injury. We established a TBBPA exposure model using common carp and primary renal cells to explore the mechanism of TBBPA-induced renal injury and the therapeutic efficacy of TPs, using histopathological examination and molecular biology techniques. This study offers novel insights into the treatment of TBBPA-induced renal injury in common carp.

2. Materials and Methods

2.1. Animals and Treatment

All experimental procedures involving animals were approved by the Institutional Animal Care and Use Committee of Northeast Agricultural University. Eighty healthy carp weighing 76.0 ± 5.4 g were obtained for this study from a nearby aquaculture facility. Each experimental group contained 20 fish (n = 20). The fish were randomly divided into four groups: Control Group (Con): fed basal diet only; TPs Group (TPs): fed basal diet supplemented with 1 g/kg TPs; TBBPA Group (TBBPA): exposed to 0.5 mg/L TBBPA in water while fed basal diet; TBBPA + TPs Group (TBBPA + TPs): fed basal diet with 1 g/kg TPs and exposed to 0.5 mg/L TBBPA in water. TPs were thoroughly mixed with the basal diet at 1 g/kg feed concentration. Fish were housed in the tanks with temperature controlled at 26 ± 2 °C and minimum dissolved oxygen levels above 7 mg/L. Throughout the 14-day experiment, fish were fed twice daily, with 30% water exchange performed every other day to maintain water quality. Subsequently, the common carp were anaesthetised with 0.02% MS-222 and euthanised. The kidneys were removed, a part of the kidneys rinsed with PBS, and fixed in 4% paraformaldehyde. The remaining kidney segments were frozen in liquid nitrogen and stored at −80 °C for later tests.

2.2. Culture and Treatment of Primary Renal Cells from Carp Kidney

Healthy carp were selected, then the body surface was first wiped with 0.1% potassium permanganate solution, followed by disinfection with 75% ethanol at the dissection site. Approximately 0.5 cm^3 of renal tissue was aseptically excised. After removal of surface blood clots and peritoneal tissue, the tissue was minced and rinsed 3–4 times with phosphate-buffered saline (PBS). The renal tissues were transferred into EP tubes, digested with 0.25% trypsin, and placed in a 28 °C water bath for 5 min with manual shaking every minute. After digestion, the trypsin solution was aspirated, and the tissues were completely dissociated with complete MI99 medium [24]. The cells were then seeded into six-well plates at densities ranging from 2×10^6 to 8×10^6 cells per well. The culture medium consisted of complete MI99 medium supplemented with 12% fetal bovine serum (Haixing Biological Technology Co., Ltd., Suzhou, China) and 1% penicillin-streptomycin (Solarbio Science & Technology Co., Ltd., Beijing, China). Cells were maintained at 28 °C in a 5% CO_2 incubator. When the cell reached 80% confluency, the cultures were divided into three groups: Control group, 60 μM TBBPA group, 2.5 μg/L TPs group, and a 60 μM TBBPA + 2.5 μg/L TPs group. After 24 h of treatment, cells were harvested for subsequent analysis.

Cell viability was assessed using a Cell Counting Kit-8 (CCK-8; Jiancheng Biotechnology Co., Ltd., Nanjing, China). Renal primary cells were seeded in 96-well plates, and complete medium supplemented with varying concentrations of TBBPA (0, 20, 40, 60, 80, 100, or 120 μM) was added to each well. The Con group received medium with 0.1% DMSO (v/v). After 48 h of incubation, the medium was discarded, and 10 μL of CCK-8 solution was added to each well before incubation at 28 °C for 2 h. The absorbance was measured at 450 nm using a microplate reader. The viability of primary renal cells could reach 80% after being treated with 60 μM TBBPA (Figure S1). Then, 60 μM TBBPA was used for subsequent experiments.

2.3. *Histological Analysis*

Fresh carp renal tissues were fixed in 4% paraformaldehyde (PFA), dehydrated through a graded ethanol series, cleared in xylene, and embedded in liquid paraffin. After solidification, the paraffin blocks were sectioned at 5 μm thickness. Following dewaxing and hydration, the sections were stained with hematoxylin and eosin (HE) [25]. The stained sections were sealed, and the morphology of the renal tissues was observed and photographed under a microscope (Olympus Optical Co., Ltd., Tokyo, Japan). Three tissue blocks per group were analysed (n = 3).

2.4. *TUNEL Analysis*

Renal tissues were rinsed with PBS before paraffin embedding. The paraffin-embedded sections were dewaxed in xylene and rehydrated through a graded ethanol series (70%, 80%, 95%, 100%), followed by incubation with proteinase K working solution at 37 °C. Subsequently, the sections were sequentially stained with terminal deoxynucleotidyl transferase (TdT) and deoxyuridine triphosphate (dUTP) solutions. Following reaction termination, the sections were washed to remove residual reagents and mounted for microscopic analysis [26]. Fluorescence microscopy (Olympus Optical Co., Ltd., Tokyo, Japan) was performed to examine the fluorescence signals in the renal tissues. Three tissue blocks per group were analysed (n = 3).

2.5. *Oxygen Radical Detection*

The renal tissue was trimmed into 1 mm^3 pieces and sectioned. Using fine-tipped forceps, the tissue was gently dissociated by rubbing against a 300-mesh nylon net, followed by PBS rinsing until complete tissue dissociation was achieved. A single-cell suspension was prepared through resuspension, then incubated with a diluted DCFH-DA probe at 37 °C for 30 min. The probe-labeled single-cell suspension was collected, washed, and resuspended. Fluorescence intensity was measured using a microplate fluorometer at an excitation wavelength of 500 nm and an emission wavelength of 525 nm [22]. For the in vitro experiments, the supernatant from TBBPA- and TPs-treated cells was discarded, and cells were washed three times with PBS. The DCFH-DA probe was prepared by 1:1000 dilution in serum-free medium, and 1 mL of the diluted probe was added to each well. Following incubation at 37 °C for 30 min, cells were washed three times with PBS and analysed by fluorescence microscopy [27]. The ROS kit was purchased from Nanjing Jiancheng Biotechnology Co., Ltd., Nanjing, China.

2.6. *AO/EB Staining*

Necrosis was observed using the acridine orange/ethidium bromide (AO/EB) double dye kit (Leagene, Biotechnology Co., Ltd., Beijing, China). The treated cells were harvested and mixed with AO/EB (acridine orange/ethidium bromide) at a 1:1 ratio in PBS to prepare a working solution (10 μg/mL). The cell suspension was then replated in 6-well plates and visualised by fluorescence microscopy (Olympus Optical Co., Ltd., Tokyo, Japan) [28]. Each group included three independent biological replicates (n = 3). Fluorescence images were acquired. Quantitative analysis was performed using ImageJ 1.53e software.

2.7. *Antioxidant Enzyme Detection*

A 10% tissue homogenate was prepared by mixing 900 mL of physiological saline with 0.1 g of sample in a homogeniser maintained in an ice bath. Collect the primary renal cells, wash them 1–2 times with PBS, then centrifuge at low speed to pellet the cells. Resuspend the cell pellet in 0.3–0.5 mL of PBS buffer, and homogenise the cells by grinding. Centrifuge at 3500 rpm for 10 min, and collect the supernatant for the chromogenic reaction. Each group included three independent biological replicates ($n = 3$). Protein concentrations in tissues and cells were determined using a BCA protein assay kit (Thermo Fisher Scientific, Waltham, MA, USA). The levels of superoxide dismutase (SOD), glutathione (GSH), malondialdehyde (MDA), and catalase (CAT) in tissues and cells were quantified using corresponding assay kits (Jiancheng Biotechnology Co., Ltd., Nanjing, China) according to the manufacturer's instructions.

2.8. Enzyme-Linked Immunosorbent Assay (ELISA)

Protein expression levels of interleukin-1β (IL-1β), interleukin-6 (IL-6), and tumour necrosis factor-α (TNF-α) were measured in both renal tissues and isolated primary renal cells using commercial ELISA kits from Andy Gene Biotechnology, China (catalogue numbers: E-43902 for IL-1β, E-53219 for IL-6, and E-43904 for TNF-α). Each experimental group included three biological replicates. All assay procedures, including sample preparation and incubation conditions, were strictly followed according to the manufacturer's instructions. The absorbance was measured at 450 nm using a Multiskan SkyHigh microplate reader (Thermo Fisher Scientific, Waltham, MA, USA).

2.9. Real-Time PCR

Renal tissues and primary renal cells were lysed with Trizol reagent (Thermo Fisher Scientific, Waltham, MA, USA) following the manufacturer's protocol for RNA extraction. Total RNA concentration was assessed by UV spectrophotometry (Thermo Fisher Scientific, Waltham, MA, USA). RNA was reverse transcribed into complementary DNA (cDNA) using a reverse transcription kit (Vazyme Biotech Co., Ltd., Nanjing, China). Target genes were denatured, annealed, and amplified in a 20 μL reaction system using a PCR 2720 Thermal Cycler (Thermo Fisher Scientific, Waltham, MA, USA) and an ABI 7500 Real-Time PCR System. Glyceraldehyde-3-phosphate dehydrogenase (GAPDH) was used as the internal reference gene, and the results were expressed as $2^{-\Delta\Delta Ct}$. Each group included three independent biological replicates ($n = 3$). Primer sequences are listed in Table 1.

Table 1. The primers used in the present study.

Gene	Forward Primer (5′→3′)	Reverse Primer (5′→3′)	Accession Number
PI3K	CGGGAAACGAGCTCAATCAT	CTCTCCAACAAATCCGCTCG	KY763989
AKT	CAACGGATGCGTCGTCTTCA	GTGTGAGTCTCCAAACCCCT	JX307852
IKB-α	GGCTACGCCAAAGACCTG	CGGACCTCGCCATTCATA	AK313421
NF-κB p65	GAAGAAGGATGTGGGAGATG	TGTTGTCGTAGATGGGCTGAG	MN167531
IL-1β	CTTCCACCCTCACAAACACATTCAAC	AATATAGCGTCCAAGGCGTTCCATC	EU047716
IL-6	CCGCATGGACTCGCAAGACG	CGGTAGTTGATGTACTCGTCCTCC	KC858890
TNF-α	TCATGGGAGTAAGGCTGGTATT	TCTTCAAAGGAATACAGGGGCT	LN593053
Caspase-3	GCTGTGCTTCGTTAGTGT	GAACCAAGAACCGCTCAT	JAEOAB010000019
Bax	ATGCGTGAATAAGGAGATGA	AGACCGAAGACCGTTACT	KJ174685
Bcl-2	GATACCGCAAGATTCCATACCC	TCCTTTCTATCTCGTCTCCAG	KJ174686
RIPK1	GGCTGCGTCGTTTGATA	GTTGGCACCCACGTTCT	MN123251
RIPK3	CAACGATGCCGTCTATGA	GAAGGAGCTGTTTGGTGTCT	OY720469
MLKL	CTGGCACAACAATCTGA	GAGACGCTGTAGAAGGAC	BC028141
GAPDH	GTTACAAGGGAGAAGTTCACCAT	CCGGTAGACTCGACTACATACAG	AJ870982

2.10. Western Blotting Assay

Proteins were extracted from the kidney and renal progenitor cells using lysis buffer, the optical density (OD) values of standards and samples were measured using a BCA Protein Assay Kit (Jiancheng Biotechnology Co., Ltd., Nanjing, China), and standard curves were generated to calculate sample protein concentrations. An appropriate SDS-polyacrylamide gel concentration was selected based on the molecular weight of the target proteins. The target bands were cut out according to the molecular weight markers. Target protein bands were excised according to molecular weight markers and transferred to nitrocellulose membranes by wet blotting at 220 mA. After transfer, the membrane was blocked with 5% skim milk for 2 h at room temperature, then incubated with primary antibody at 4 °C overnight. Subsequently, the membrane was incubated with diluted secondary antibodies (goat anti-rabbit IgG and goat anti-mouse IgG) for 2 h at room temperature in the dark for signal detection [29]. Each group included three independent biological replicates ($n = 3$). Antibodies used for Western blot analysis are listed in Table 2.

Table 2. Antibodies required for Western blot.

Name	Cat No.	Company	Dilution Times
p-PI3K	AF5905	Beyotime Biotechnology Co., Ltd., Shanghai, China	1:2000
P-AKT	AA329	Beyotime Biotechnology Co., Ltd., Shanghai, China	1:1000
NF-κB p65	AF1234	Beyotime Biotechnology Co., Ltd., Shanghai, China	1:1000
P-NF-κB p65	AF5875	Beyotime Biotechnology Co., Ltd., Shanghai, China	1:1000
IKB-α	AF1282	Beyotime Biotechnology Co., Ltd., Shanghai, China	1:1000
p-IKB-α	AF1870	Beyotime Biotechnology Co., Ltd., Shanghai, China	1:1000
Bax	WL01637	Wanlei Life Sciences Co., Ltd., Shenyang, China	1:500
Bcl-2	WL01556	Wanlei Life Sciences Co., Ltd., Shenyang, China	1:500
Caspase-3	WL04004	Wanlei Life Sciences Co., Ltd., Shenyang, China	1:500
RIPK3	47928T	Cell Signaling Technology, Inc., Boston, MA, USA	1:1000
p-RIPK3	93654T	Cell Signaling Technology, Inc., Boston, MA, USA	1:1000
MLKL	14993T	Cell Signaling Technology, Inc., Boston, MA, USA	1:1000
p-MLKL	3733T	Cell Signaling Technology, Inc., Boston, MA, USA	1:1000
GAPDH	WL01114	Wanlei Life Sciences Co., Ltd., Shenyang, China	1:1000

2.11. Statistical Analysis

Data were analysed using GraphPad Prism 10 software. One-way analysis of variance (ANOVA) was used to calculate the difference between the values. All experiments included triplicate biological replicates ($n = 3$). Statistical significance was determined as follows: (*, $p < 0.05$; **, $p < 0.01$; ***, $p < 0.001$; ****, $p < 0.0001$).

3. Results

3.1. TPs Mitigated TBBPA-Induced Renal Injury

Histopathological analysis demonstrated that TBBPA exposure induced renal damage compared to the Con group (Figure 1A,B), characterised by renal tubular oedema and rupture, tissue congestion, and inflammatory cell infiltration. After TPs intervention, although residual congestion persisted in renal tissues, renal tubular morphology showed improvement (Figure 1C). No morphological alterations were observed in the TPs group (Figure 1D).

3.2. TPs Mitigated the Oxidative Stress Induced by TBBPA in Renal Tissue

Compared to the Con group, the TBBPA group exhibited significantly higher ROS levels in carp renal tissues ($p < 0.0001$). TPs treatment reduced ROS accumulation (Figure 2A). Fluorescence was used to detect the expression of reactive oxygen species (ROS) in primary renal cells from carp kidney. TBBPA exposure elevated oxidative stress levels in cells. The green fluorescence intensity was reduced by TPs treatment (Figure 2E–H). The results showed that MDA levels were significantly higher ($p < 0.0001$) in the TBBPA group than in the Con group in tissue and cells, but TPs treatment effectively reduced them ($p < 0.001$) (Figure 2B,I). Antioxidant enzyme activity results are presented in Figure 2C,D,J,K. Compared with the Con group, the TBBPA group exhibited reduced CAT and SOD levels, whereas TPs treatment significantly enhanced ($p < 0.001$) their activities.

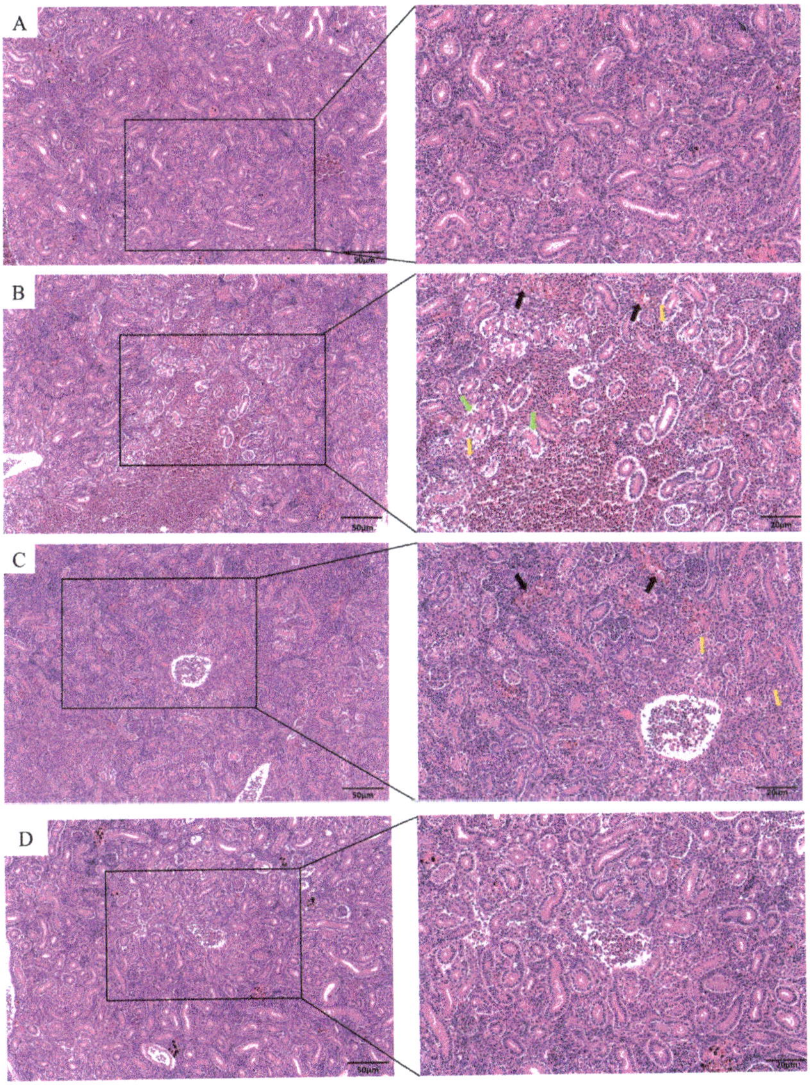

Figure 1. Effects of TBBPA exposure and TPs intervention on the morphology of carp kidney. (**A**) HE staining of kidney tissue in the Con group. (**B**) HE staining of kidney tissue in the TBBPA group. (**C**) HE staining of kidney tissue in the TBBPA + TPs group. (**D**) HE staining of kidney tissue in the TPs group. The black arrows in the figure indicate congestion of the renal tissue, the yellow arrows indicate inflammatory cell infiltration, and the green arrows indicate oedema of the renal tubules. The left figure is magnified 20×, and the right figure is magnified 40×. Each group included three independent biological replicates ($n = 3$). The data are expressed as mean ± SEM and analysed using one-way analysis of variance (ANOVA).

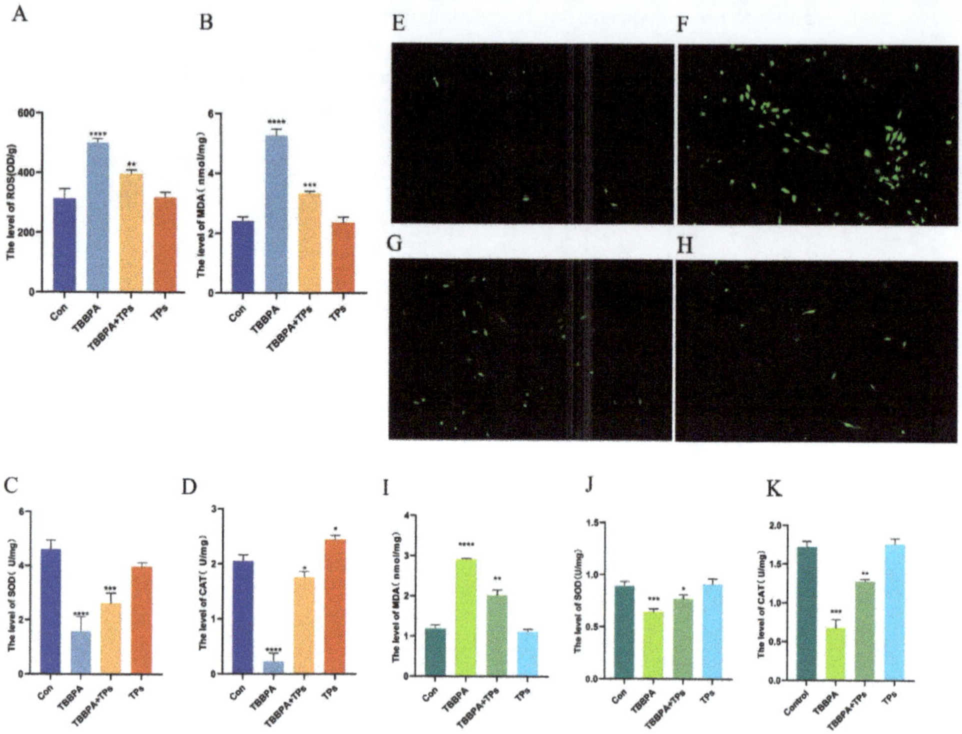

Figure 2. Effects of TBBPA exposure and TPs intervention on renal oxidative stress of common carp. (**A**) Changes in ROS levels in the kidney. (**B**) MDA content was detected in kidney tissue. (**C**) SOD content was detected in the kidney. (**D**) CAT content was detected in the kidney. (**E–H**) Changes in the content of ROS in cells, magnification 10×. (**I**) Detection of MDA content in primary renal cells. (**J**) Detection of SOD content in primary renal cells. (**K**) Detection of CAT content in primary renal cells. Each group consisted of three independent biological replicates ($n = 3$). Data are expressed as mean ± SEM. Analysis was performed using one-way ANOVA. *, $p < 0.05$; **, $p < 0.01$; ***, $p < 0.001$; ****, $p < 0.0001$.

3.3. TPs Mitigated TBBPA-Induced Renal Inflammatory Injury

To further evaluate renal inflammatory responses, we examined the expression of inflammatory factors IL-1β, IL-6, and TNF-α in both tissue and cells. The results demonstrated that the mRNA levels of *IL-1β*, *IL-6*, and *TNF-α* remained comparable to those of the Con group when TPs were given alone, whereas the mRNA levels of *IL-1β*, *IL-6*, and *TNF-α* were significantly elevated ($p < 0.0001$) by TBBPA exposure (Figure 3A–C,G–I). As shown in Figure 3D–L, the protein levels of inflammatory factors followed the same trend as their mRNA counterparts. The levels of inflammatory factors were significantly increased ($p < 0.0001$) in the TBBPA group. The levels of inflammatory factors were decreased after the TPs intervention.

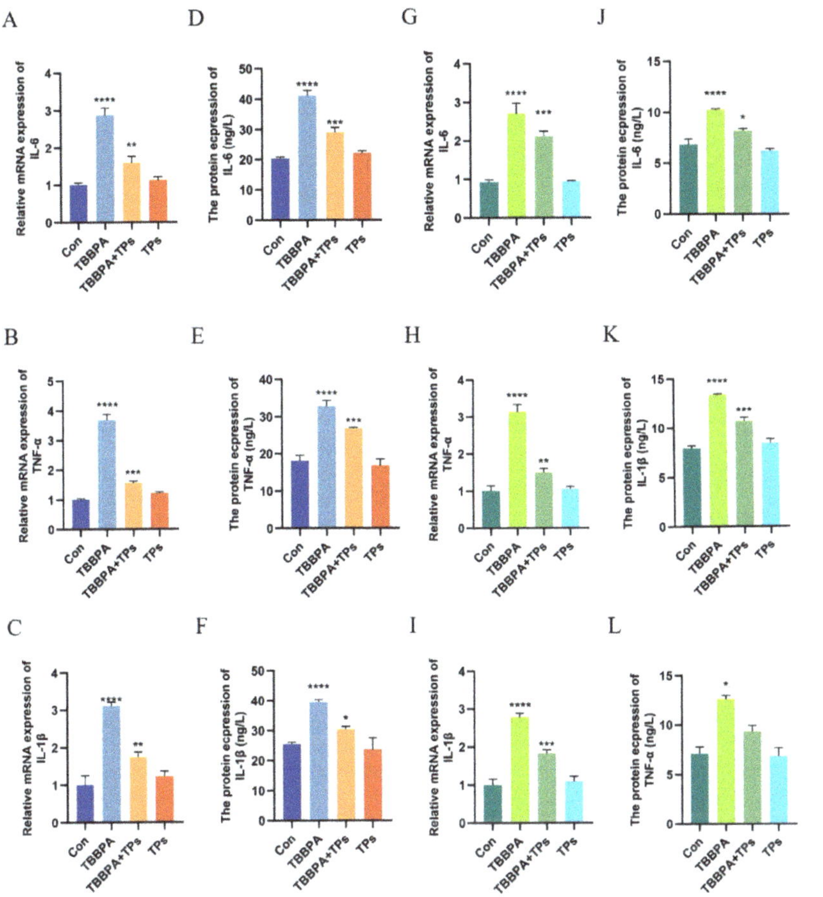

Figure 3. Effects of TBBPA exposure and TPs intervention on the levels of inflammatory factors in renal tissue. (**A–C**) The mRNA levels of inflammatory factors in renal tissue. (**D–F**) Protein levels of inflammatory factors in renal tissue. (**G–I**) Expression levels of mRNA of inflammatory factors in primary renal cells. (**J–L**) Expression levels of the protein of inflammatory factors in primary renal cells. Each group included three independent biological replicates ($n = 3$). The data are expressed as mean ± SEM. Analysis was performed using one-way ANOVA. *, $p < 0.05$; **, $p < 0.01$; ***, $p < 0.001$; ****, $p < 0.0001$.

3.4. TPs Mitigated TBBPA-Induced Activation of the PI3K/AKT/NF-κB Pathway

As shown in Figure 4A–F, TBBPA exposure significantly increased ($p < 0.0001$) both PI3K and AKT levels at the mRNA and protein levels in tissues and cells compared with the Con group. p-PI3K and p-AKT were also markedly increased ($p < 0.0001$) in the TBBPA group. TPs treatment significantly reduced ($p < 0.0001$) these phosphorylation levels. TBBPA exposure significantly upregulated ($p < 0.0001$) phosphorylated IKB-α (p-IKB-α) and phosphorylated NF-κB (p-NF-κB) at both the transcriptional and translational levels (Figure 4A–F) compared to the Con group. TPs treatment significantly decreased ($p < 0.001$) the protein levels of both p-IKB-α and p-NF-κB.

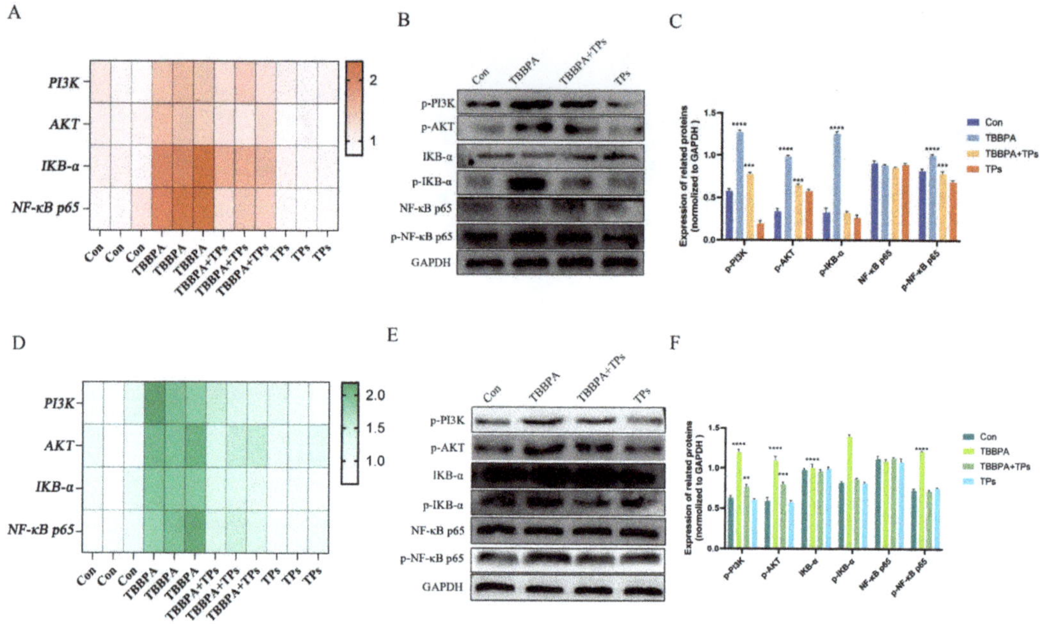

Figure 4. Effects of TBBPA exposure and TPs intervention on the expression of the PI3K/AKT-NF-κB pathway. (**A**) The mRNA levels of key proteins in the PI3K/AKT/NF-κB pathway in renal tissue. (**B**,**C**) Expression levels of key proteins in the PI3K/AKT/NF-κB pathway in renal tissue. (**D**) The mRNA levels of key proteins in the PI3K/AKT/NF-κB pathway in primary renal cells. (**E**,**F**) Levels of key proteins in the PI3K/AKT/NF-κB pathway in primary renal cells. Each group included three independent biological replicates ($n = 3$). Data are expressed as mean ± SEM. Analysis was performed using one-way analysis of variance (ANOVA). **, $p < 0.01$; ***, $p < 0.001$; ****, $p < 0.0001$.

3.5. TPs Reduced TBBPA-Induced Renal Cell Apoptosis

TUNEL results demonstrated that the number of apoptotic cells in kidney tissues was significantly elevated ($p < 0.001$) in the TBBPA group compared with the Con group. This apoptotic effect was significantly reduced ($p < 0.0001$) after TPs intervention (Figure 5A,B). As shown in Figure 5C,F, the mRNA levels of *Caspase-3* and *BAX* were significantly increased ($p < 0.001$) in the renal tissue and primary renal cells after TBBPA exposure, while the mRNA levels of the anti-apoptotic gene *BCL-2* were significantly decreased ($p < 0.0001$). TPs intervention decreased ($p < 0.05$) *Caspase-3* and *BAX* mRNA levels while increasing *BCL-2* expression. Western blot results similarly showed that the protein levels of Caspase-3 and BAX were significantly increased ($p < 0.0001$) after TBBPA exposure (Figure 5D,E,G,H). At the same time, those of BCL-2 were significantly decreased ($p < 0.0001$). Caspase-3 and BAX protein levels were significantly decreased ($p < 0.001$), downregulated, and BCL-2 protein levels were significantly increased, upregulated, after TPs intervention (Figure 5D,E,G,H).

3.6. TPs Mitigated TBBPA-Induced Renal Cell Necrosis

Figure 6A,E demonstrate that TBBPA exposure significantly elevated ($p < 0.001$) the mRNA levels of receptor-interacting protein kinase 1 (*RIPK1*), receptor-interacting protein kinase 3 (*RIPK3*), and mixed lineage kinase domain-like protein (*MLKL*) in tissues and cells, while TPs intervention effectively reversed this effect. The results of acridine orange and ethidium bromide (AO/EB) staining are shown in Figure 6D. Compared with the Con group, the number of apoptotic and necrotic cells after TBBPA exposure was increased, and the number of apoptotic and necrotic cells after TPs intervention was significantly decreased. Western blot analysis revealed that the protein levels of phosphorylated RIPK3 (p-RIPK3) and phosphorylated MLKL (p-MLKL) were markedly upregulated ($p < 0.0001$) upon TBBPA treatment, which were subsequently attenuated by TPs administration (Figure 6B,C,F,G).

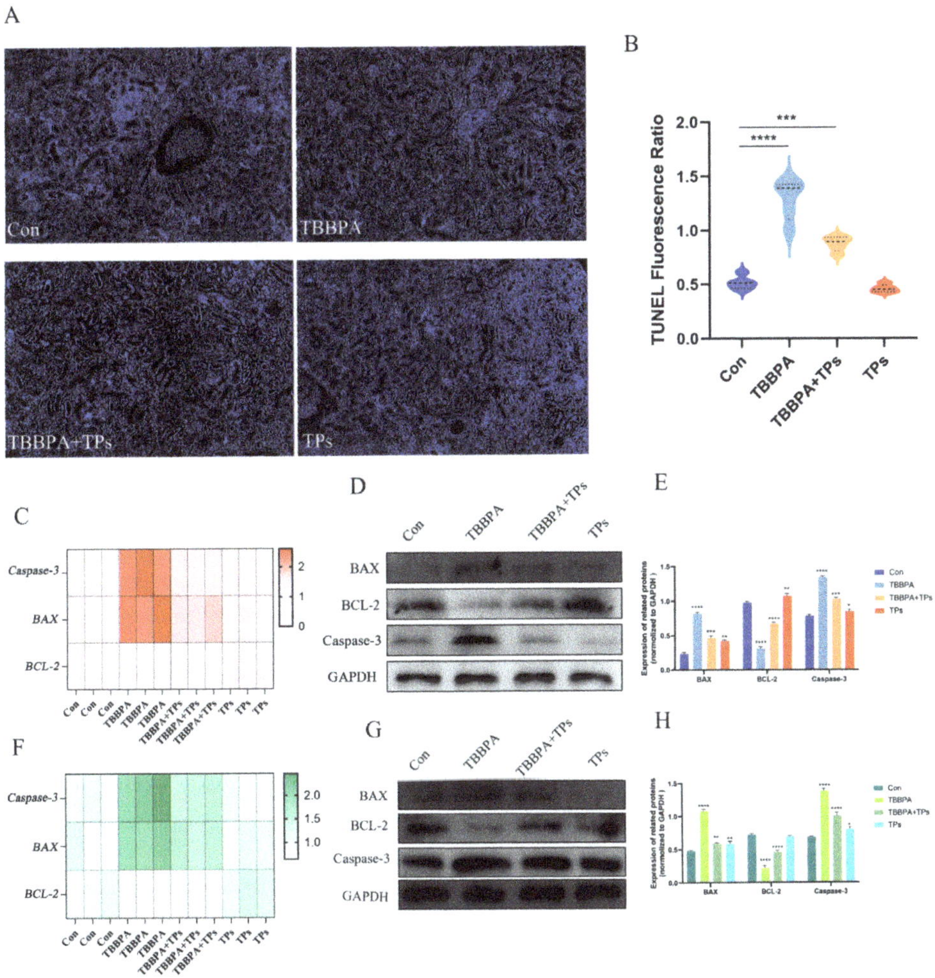

Figure 5. Effects of TBBPA exposure and TPs intervention on renal apoptosis of common carp. (**A**) TUNEL results of kidney tissue; green represents apoptosis. (**B**) Tunnel fluorescence ratio. (**C**) The mRNA expression of *Caspase-3*, *BAX*, and *BCL-2* genes in the kidney. (**D**,**E**) Protein expression levels of Caspase-3, BAX, and BCL-2 in the kidney. (**F**) The mRNA expression of Caspase-3, BAX, and BCL-2 in primary renal cells. (**G**,**H**) Protein expression of the *Caspase-3* gene and *BAX* and *BCL-2* in primary renal cells. Each group included three independent biological replicates (n = 3). Data are expressed as mean ± SEM. Analysis was performed using one-way ANOVA. *, $p < 0.05$; **, $p < 0.01$; ***, $p < 0.001$; ****, $p < 0.0001$.

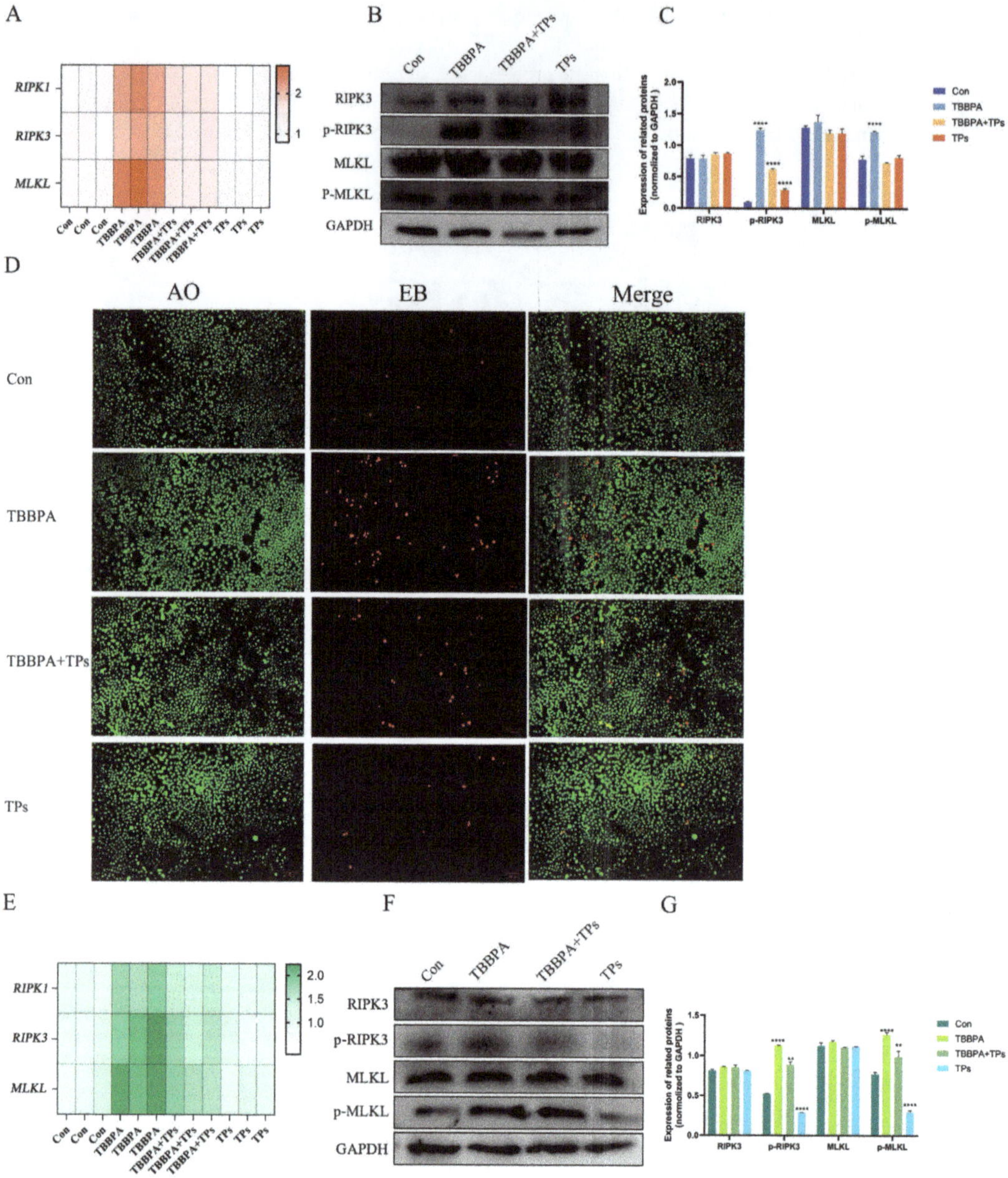

Figure 6. Effects of TBBPA exposure and TPs intervention on renal necrosis of carp. (**A**) The mRNA expression of *RIPK1*, *RIPK3*, and *MLKL* genes in the kidney. (**B**,**C**) Expression levels of RIPK1, RIPK3, and MLKL proteins in the kidney. (**D**) AO/EB results. Orange represents apoptotic cells, and red represents necrotic cells. Magnification 10×. (**E**) The mRNA expression of *RIPK1*, *RIPK3*, and *MLKL* genes in renal primary cells. (**F**,**G**) Protein expression levels of RIPK1, RIPK3, and MLKL in primary renal cells. Each group included three independent biological replicates ($n = 3$). Data are expressed as mean ± SEM. Analysis was performed using one-way analysis of variance (ANOVA). **, $p < 0.01$; ****, $p < 0.0001$.

4. Discussion

TBBPA is the most widely produced brominated flame retardant globally and is a widespread persistent organic pollutant in the environment [5]. Studies have demonstrated that TBBPA adversely affects the survival, reproduction, and development of various aquatic organisms [30]. Tea polyphenols are a general term for polyphenols in tea, which exhibit pharmacological effects such as antioxidant, antiviral, and antibacterial properties [31]. Tea polyphenols (TPs) not only enhance the antioxidant capacity of common carp but also modulate the gut microbiota composition, thereby preventing intestinal barrier dysfunction. Specifically, TPs alleviate *Aeromonas hydrophila*-induced intestinal barrier damage in grass carp by inhibiting the RhoA/ROCK signalling pathway, and upregulating the expression of tight junction and adherens junction proteins [32]. Additionally, TPs enhance non-specific immunity and exert anti-inflammatory effects in koi carp by suppressing pro-inflammatory cytokines (IL-1β, IL-6) via the NF-κB pathway, while altering gut microbial structure (increasing Proteobacteria and reducing Fusobacteria abundance) [33]. The aim of this study was to investigate the protective effects of tea polyphenols on TBBPA-induced renal inflammation, apoptosis, and necroptosis. The results indicated that TBBPA induced inflammation, apoptosis, and necroptosis in kidney tissues and primary renal cells of common carp, whereas TPs intervention mitigated these effects.

The common carp is a freshwater fish of the genus Cyprinus. The kidneys of freshwater fish play a pivotal role in maintaining water and electrolyte homeostasis. In freshwater environments, fish excrete excess water while retaining salts via renal filtration and active tubular reabsorption, particularly of sodium chloride, to adapt to hypoosmotic conditions [34,35]. This process relies on efficient glomerular filtration and tubular reabsorption mechanisms. At the molecular level, TBBPA may interfere with renal excretory function by reacting with biotransformation enzymes such as glutathione, reducing its bioavailability [36]. Freshwater fish kidneys regulate acid-base balance and ammonia excretion, primarily through NH_4^+ elimination and H^+/HCO_3^- reabsorption. However, ammonia imbalance induces renal oxidative stress and structural damage. Zhang et al. [37] demonstrated that ammonia exposure upregulates CPS I expression in yellow catfish kidneys, activating glutamine synthesis while still causing oxidative stress and immunosuppression. Dietary supplementation with 5% TCE (Thalassodendron ciliatum extract) enhances antioxidant capacity in tilapia, mitigates ammonia-induced oxidative stress, and may provide renal protection [38].

An imbalance between oxidation and antioxidant defence systems leads to excessive ROS production. This triggers oxidative stress, which ultimately causes tissue damage. Zhang et al. demonstrated that TBBPA induces ROS overproduction and upregulates antioxidant enzymes. Malondialdehyde (MDA), an oxidative stress marker, induced oxidative stress in L02 cells [39]. The results showed that elevated concentrations of ROS and MDA were observed in kidney cells after TBBPA exposure. SOD acts as the primary defence against ROS by detoxifying superoxide anions. CAT, on the other hand, scavenges hydrogen peroxide, which further protects the cells from the damage caused by ROS [40]. TBBPA intervention significantly reduces the expression of SOD and CAT in mouse skeletal muscle and C2C12 cells [41]. Consistently, this study showed that the expression levels of SOD and CAT were significantly decreased in the kidney and primary renal cells of common carp after TBBPA treatment. TPs are potent antioxidants that scavenge free radicals and regulate cellular ROS levels, counteracting oxidative damage [42]. Studies have demonstrated that TPs increase serum CAT and SOD while decreasing MDA production in rats [43]. These findings indicate that tea polyphenols can alleviate oxidative stress and enhance endogenous antioxidant defences. The results of this study demonstrate that dietary

TPs can scavenge excess ROS and reduce MDA levels, thereby enhancing the antioxidant capacity in the kidneys of common carp. The present findings are in concordance with the well-documented antioxidant properties of TPs reported in prior studies.

It has been shown that ROS are important regulatory molecules of the PI3K/AKT pathway and have a direct effect on AKT under oxidative stress [44]. The PI3K/AKT pathway serves as an important upstream regulator of the NF-κB signalling pathway [45]. PI3K activation converts phosphatidylinositol 4,5-bisphosphate (PIP2) to phosphatidylinositol 3,4,5-trisphosphate (PIP3), subsequently activating AKT. This activation leads to the release of inhibited NF-κB by phosphorylating IKB-α, which in turn initiates the transcription of pro-inflammatory genes [46]. In this study, the PI3K/AKT/NF-κB pathway was activated, and the expression levels of IL-1β, IL-6, and TNF-α were increased in common carp renal tissue and primary renal cells after TBBPA exposure. These results demonstrate that TBBPA activates the PI3K/AKT/NF-κB signalling pathway, thereby upregulating inflammatory factors and triggering inflammatory responses. TPs are considered natural compounds with anti-inflammatory potential. Previous studies have shown that green tea polyphenols can mitigate TBBPA-induced oxidative stress and alleviate inflammatory responses by inhibiting NF-κB activation and reducing lung injury in mice [47]. The current study demonstrated that TPs inhibited PI3K/AKT/NF-κB pathway activation and reduced production of inflammatory factors (IL-1β, IL-6, and TNF-α).

The pro-apoptotic proteins Caspase-3 and Bax and the anti-apoptotic protein Bcl-2 can regulate apoptosis. Xu et al. demonstrated that TBBPA induces gastric mucosal apoptosis by upregulating Caspase-3 through ROS-mediated NF-κB activation [8]. This study demonstrated that TBBPA exposure upregulated Caspase-3 and BAX while downregulating BCL-2 in common carp renal tissues and primary renal cells, indicating TBBPA activates apoptotic pathways through these molecular changes. Ma et al. demonstrated that against TBBPA-induced damage, TPs could enhance the antioxidant capacity and decrease the expression of pro-apoptotic genes in the gills of grass carp, thereby inhibiting excessive apoptosis and alleviating inflammation [33]. Consistent with these findings, our study demonstrated that TPs protected common carp renal tissues by attenuating apoptosis. TPs also reduce apoptosis-related genes (*Caspase-3*, *BAX*, *BCL-2*) and attenuate apoptosis in renal tissues and primary renal cells.

Cell death is a crucial process in organism development and homeostasis, with apoptosis and necrosis being the two main modes of cell death [4]. RIPK3 and its substrate MLKL are considered core cellular regulators of programmed necrosis, while TNF family cytokines such as TNF-α can mediate activation of the RIPK3/MLKL pathway [48]. Xu et al. showed that TBBPA exposure significantly increased the expression levels of necroptosis-related genes RIPK3 and MLKL, leading to gastric mucosal necrosis in mice. The results of this study showed that the expression levels of necrosis-associated genes RIPK1, RIPK3, and MLKL were significantly elevated in common carp kidneys and renal progenitor cells after TBBPA exposure, indicating that TBBPA exposure activates the RIPK3/MLKL signalling pathway to induce necrosis in common carp kidneys.

5. Conclusions

Dietary supplementation with TPs suppresses TBBPA-induced excessive ROS production, thereby alleviating oxidative stress in the kidney and primary renal cells of common carp. This inhibition further blocks the PI3K/AKT/NF-κB signaling pathway, mitigating inflammatory injury, apoptosis, and necroptosis. The potential synergistic effects between tea polyphenols and other antioxidants (e.g., vitamin E, selenium) warrant further investigation.

Supplementary Materials: The following supporting information can be downloaded at: https://www.mdpi.com/article/10.3390/ani15152307/s1, Figure S1: Effects of TBBPA on the viability of primary renal cells. * $p \leq 0.05$; ** $p \leq 0.01$; *** $p \leq 0.001$. $p < 0.05$ indicates significant differences.

Author Contributions: F.H.: Conceptualisation, Methodology, Investigation, Data curation, Writing—original draft. R.X.: Investigation, Methodology. H.W.: Validation, Formal analysis. X.G.: Methodology, Formal analysis, Writing—review & editing; M.G.: Resources; Supervision, Writing—review & editing. All authors have read and agreed to the published version of the manuscript.

Funding: This work was supported by the Natural Science Foundation of Heilongjiang [LH2023C028].

Institutional Review Board Statement: The study was approved by the Ethics Committee of Northeast Agriculture University (NEAUEC2025 03 65), on [12 March 2025].

Informed Consent Statement: Not applicable.

Data Availability Statement: Data is contained within the article or Supplementary Material.

Acknowledgments: The authors extend their sincere thanks to the members of the veterinary internal medicine laboratory and key Laboratory for Laboratory Animals at the College of Veterinary Medicine, Northeast Agricultural University.

Conflicts of Interest: The authors declare no conflicts of interest.

References

1. Song, Y.; Zhang, W. Balancing Growth and Sustainability in China's Carp Aquaculture: Practices, Policies, and Sustainability Pathways. *Sustainability* **2025**, *17*, 5593. [CrossRef]
2. Cao, J.; Chen, J.; Xie, L.; Wang, J.; Feng, C.; Song, J. Protective properties of sesamin against fluoride-induced oxidative stress and apoptosis in kidney of carp (Cyprinus carpio) via JNK signaling pathway. *Aquat. Toxicol.* **2015**, *167*, 180–190. [CrossRef] [PubMed]
3. Milla, S.; Wang, N.; Mandiki, S.N.; Kestemont, P. Corticosteroids: Friends or foes of teleost fish reproduction? *Comp. Biochem. Physiol. Part A: Mol. Integr. Physiol.* **2009**, *153*, 242–251. [CrossRef]
4. Serradimigni, R.; Rojas, A.; Leong, C.; Pal, U.; Bryan, M.; Sharma, S.; Dasgupta, S. Flame retardant tetrabromobisphenol A (TBBPA) disrupts histone acetylation during zebrafish maternal-to-zygotic transition. *J. Hazard. Mater.* **2024**, *480*, 135845. [CrossRef]
5. Abdallah, M.A.E. Environmental occurrence, analysis and human exposure to the flame retardant tetrabromobisphenol-A (TBBP-A)-A review. *Environ. Int.* **2016**, *94*, 235–250. [CrossRef]
6. Ronisz, D.; Finne, E.F.; Karlsson, H.; Förlin, L. Effects of the brominated flame retardants hexabromocyclododecane (HBCDD), and tetrabromobisphenol A (TBBPA), on hepatic enzymes and other biomarkers in juvenile rainbow trout and feral eelpout. *Aquat. Toxicol.* **2004**, *69*, 229–245. [CrossRef]
7. Anh, H.Q.; Minh, T.H.; Minh, P.D.; Nhat, T.M.; Minh, N.L.H.; Minh, T.B.; Takahashi, S. Analysis and Pollution Assessment of Brominated Flame Retardants (PBDEs, DBDPE, and BTPBE) in Settled Dust from E-waste and Vehicle Processing Areas in Northern Vietnam. *VNU J. Sci. Nat. Sci. Technol.* **2023**, *39*, 33–40. [CrossRef]
8. Xu, S.; Sun, X.; Wu, J.; Li, K.; Li, X.; Zhang, Y.; Gao, X.J. TBBPA causes inflammation and cell death via the ROS/NF-κB pathway in the gastric mucosa. *Ecotoxicol. Environ. Saf.* **2023**, *262*, 115320. [CrossRef]
9. Okeke, E.S.; Huang, B.; Mao, G.; Chen, Y.; Zhengjia, Z.; Qian, X.; Wu, X.; Feng, W. Review of the environmental occurrence, analytical techniques, degradation and toxicity of TBBPA and its derivatives. *Environ. Res.* **2022**, *206*, 112594. [CrossRef]
10. Liu, K.; Li, J.; Yan, S.; Zhang, W.; Li, Y.; Han, D. A review of status of tetrabromobisphenol A (TBBPA) in China. *Chemosphere* **2016**, *148*, 8–20. [CrossRef]
11. Birnbaum, L.S.; Staskal, D.F. Brominated flame retardants: Cause for concern? *Env. Health Perspect* **2004**, *112*, 9–17. [CrossRef] [PubMed]
12. Song, M.; Liang, D.; Liang, Y.; Chen, M.; Wang, F.; Wang, H.; Jiang, G. Assessing developmental toxicity and estrogenic activity of halogenated bisphenol A on zebrafish (Danio rerio). *Chemosphere* **2014**, *112*, 275–281. [CrossRef] [PubMed]
13. Yang, S.; Wang, S.; Liu, H.; Yan, Z. Tetrabromobisphenol A: Tissue distribution in fish, and seasonal variation in water and sediment of Lake Chaohu, China. *Environ. Sci. Pollut. Res.* **2012**, *19*, 4090–4096. [CrossRef]
14. Qian, M.; Geng, Y.; Wang, J.-J.; Wang, H.-R.; Luo, J.-L.; Gao, X.-J. TBBPA caused multiple intestinal injuries via ROS/NF-κB signal in common carp. *Aquat. Toxicol.* **2025**, *279*, 107190. [CrossRef] [PubMed]
15. Han, D.; Yang, N.; Liu, H.; Yao, Y.; Xu, S. TBBPA causes apoptosis in grass carp hepatocytes involving destroyed ER-mitochondrial function. *Chemosphere* **2023**, *341*, 139974. [CrossRef]

16. Fukuda, N.; Ito, Y.; Yamaguchi, M.; Mitumori, K.; Koizumi, M.; Hasegawa, R.; Kamata, E.; Ema, M. Unexpected nephrotoxicity induced by tetrabromobisphenol A in newborn rats. *Toxicol. Lett.* **2004**, *150*, 145–155. [CrossRef]
17. Tada, Y.; Fujitani, T.; Yano, N.; Takahashi, H.; Yuzawa, K.; Ando, H.; Kubo, Y.; Nagasawa, A.; Ogata, A.; Kamimura, H. Effects of tetrabromobisphenol A, brominated flame retardant, in ICR mice after prenatal and postnatal exposure. *Food Chem. Toxicol.* **2006**, *44*, 1408–1413. [CrossRef]
18. Liao, Y.; Wang, Y.; Lin, Y.; Xiao, Y.; Mohan, M.; Jaman, R.; Dong, H.; Zhu, J.; Li, X.; Zhang, C.; et al. Molecular mechanisms of tetrabromobisphenol A (TBBPA) toxicity: Insights from various biological systems. *Ecotoxicol. Environ. Saf.* **2024**, *288*, 117418. [CrossRef]
19. Shi, X.; Zhu, W.; Chen, T.; Cui, W.; Li, X.; Xu, S. Paraquat induces apoptosis, programmed necrosis, and immune dysfunction in CIK cells via the PTEN/PI3K/AKT axis. *Fish Shellfish Immunol.* **2022**, *130*, 309–316. [CrossRef]
20. Lihui, X.; Jinming, G.; Yalin, G.; Hemeng, W.; Hao, W.; Ying, C. Albicanol inhibits the toxicity of profenofos to grass carp hepatocytes cells through the ROS/PTEN/PI3K/AKT axis. *Fish Shellfish Immunol.* **2022**, *120*, 325–336. [CrossRef]
21. Frei, B.; Higdon, J.V. Antioxidant Activity of Tea Polyphenols In Vivo: Evidence from Animal Studies. *J. Nutr.* **2003**, *133*, 3275S–3284S. [CrossRef]
22. Xu, R.; Han, F.X.; Wang, H.R.; Wang, J.J.; Cai, Z.L.; Guo, M.Y. Tea polyphenols alleviate TBBPA-induced inflammation, ferroptosis and apoptosis via TLR4/NF-κB pathway in carp gills. *Fish Shellfish Immunol.* **2024**, *146*, 109382. [CrossRef]
23. Zhao, X.; Shi, X.; Liu, Q.; Li, X. Tea polyphenols alleviates acetochlor-induced apoptosis and necroptosis via ROS/MAPK/NF-κB signaling in Ctenopharyngodon idellus kidney cells. *Aquat. Toxicol.* **2022**, *246*, 106153. [CrossRef] [PubMed]
24. Mommsen, T.P.; Moon, T.W.; Walsh, P.J. Chapter 30—Hepatocytes: Isolation, maintenance and utilization. In *Biochemistry and Molecular Biology of Fishes*; Hochachka, P.W., Mommsen, T.P., Eds.; Elsevier: Amsterdam, The Netherlands, 1994; pp. 355–373.
25. Qian, M.; Yang, J.; Xue, Y.; Wu, J.; Li, Z.; Luo, J.; Zhao, B.; Gao, X. Tea Polyphenol Protects the Immune Barrier and Inhibits TLR2/NF-κB/MLCK Signal Activation to Prevent Inflammatory Injury in the Intestines of Common Carp (Cyprinus carpio L.). *Animals* **2025**, *15*, 387. [CrossRef] [PubMed]
26. Wang, F.; Zhang, Q.; Cui, J.; Bao, B.; Deng, X.; Liu, L.; Guo, M.-Y. Polystyrene microplastics induce endoplasmic reticulum stress, apoptosis and inflammation by disrupting the gut microbiota in carp intestines. *Environ. Pollut.* **2023**, *323*, 121233. [CrossRef] [PubMed]
27. Cui, J.; Xu, T.; Lv, H.; Guo, M.-Y. Zinc deficiency causes oxidative stress, endoplasmic reticulum stress, apoptosis and inflammation in hepatocytes in grass carp. *Fish Shellfish Immunol.* **2023**, *139*, 108905. [CrossRef]
28. Zhang, Q.; Wang, F.; Xu, S.; Cui, J.; Li, K.; Shiwen, X.; Guo, M.-Y. Polystyrene microplastics induce myocardial inflammation and cell death via the TLR4/NF-κB pathway in carp. *Fish Shellfish Immunol.* **2023**, *135*, 108690. [CrossRef]
29. Xu, T.; Cui, J.; Xu, R.; Cao, J.; Guo, M.-Y. Microplastics induced inflammation and apoptosis via ferroptosis and the NF-κB pathway in carp. *Aquat. Toxicol.* **2023**, *262*, 106659. [CrossRef]
30. Osako, M.; Kim, Y.-J.; Sakai, S.-I. Leaching of brominated flame retardants in leachate from landfills in Japan. *Chemosphere* **2004**, *57*, 1571–1579. [CrossRef]
31. Zhao, Z.; Zhao, F.; Cairang, Z.; Zhou, Z.; Du, Q.; Wang, J.; Zhao, F.; Wang, Q.; Li, Z.; Zhang, X. Role of dietary tea polyphenols on growth performance and gut health benefits in juvenile hybrid sturgeon (Acipenser baerii ♀× A. schrenckii ♂). *Fish Shellfish Immunol.* **2023**, *139*, 108911. [CrossRef]
32. Zhang, R.; Liu, L.L.; Wang, X.W.; Guo, C.Y.; Zhu, H. Dietary tea polyphenols induce changes in immune response and intestinal microbiota in Koi carp, cryprinus carpio. *Aquaculture* **2020**, *516*, 734636. [CrossRef]
33. Ma, Y.-B.; Jiang, W.-D.; Wu, P.; Liu, Y.; Jiang, J.; Kuang, S.-Y.; Tang, L.; Zhou, X.-Q.; Feng, L. Tea polyphenol alleviate Aeromonas hydrophila—induced intestinal physical barrier damage in grass carp (Ctenopharyngodon idella). *Aquaculture* **2021**, *544*, 737067. [CrossRef]
34. Wang, S.; Zhao, A.; Wang, B.; Zhang, Y.; Han, Q.; Yang, E.; He, H.; Wei, C.; Yang, Y.; Xu, J.; et al. Assessing the human health risks from organophosphate esters: Exposure assessment via wild freshwater fish consumption. *Food Chem.* **2025**, *492*, 145339. [CrossRef] [PubMed]
35. Sharma, G.; Chadha, P. Toxic effects of aniline in liver, gills and kidney of freshwater fish Channa punctatus after acute exposure. *Comp. Biochem. Physiol. Part C Toxicol. Pharmacol.* **2024**, *281*, 109916. [CrossRef]
36. Zhou, H.; Yin, N.; Faiola, F. Tetrabromobisphenol A (TBBPA): A controversial environmental pollutant. *J. Environ. Sci.* **2020**, *97*, 54–66. [CrossRef]
37. Zhang, M.; Wang, S.; Sun, L.; Gan, L.; Lin, Y.; Shao, J.; Jiang, H.; Li, M. Ammonia induces changes in carbamoyl phosphate synthetase I and its regulation of glutamine synthesis and urea cycle in yellow catfish Pelteobagrus fulvidraco. *Fish Shellfish Immunol.* **2022**, *120*, 242–251. [CrossRef]

38. Naiel, M.A.E.; El-Naby, A.S.A.; Samir, F.; Negm, S.S. Effects of dietary Thalassodendron Ciliatum supplementation on biochemical-immunological, antioxidant and growth indices of Oreochromis niloticus exposed to ammonia toxicity. *Aquaculture* **2024**, *585*, 740702. [CrossRef]
39. Zhang, Y.; Wang, X.; Chen, C.; An, J.; Shang, Y.; Li, H.; Xia, H.; Yu, J.; Wang, C.; Liu, Y.; et al. Regulation of TBBPA-induced oxidative stress on mitochondrial apoptosis in L02 cells through the Nrf2 signaling pathway. *Chemosphere* **2019**, *226*, 463–471. [CrossRef]
40. Wang, Y.; Branicky, R.; Noë, A.; Hekimi, S. Superoxide dismutases: Dual roles in controlling ROS damage and regulating ROS signaling. *J. Cell. Biol.* **2018**, *217*, 1915–1928. [CrossRef]
41. Zhang, Q.; Wang, S.; Wang, F.; Guo, M.; Xu, S. TBBPA induces inflammation, apoptosis, and necrosis of skeletal muscle in mice through the ROS/Nrf2/TNF-α signaling pathway. *Environ. Pollut.* **2023**, *317*, 120745. [CrossRef]
42. Yan, Z.; Zhong, Y.; Duan, Y.; Chen, Q.; Li, F. Antioxidant mechanism of tea polyphenols and its impact on health benefits. *Anim. Nutr.* **2020**, *6*, 115–123. [CrossRef] [PubMed]
43. Ahmed, N.A.; Radwan, N.M.; Ezz, H.S.A.; Salama, N.A. The antioxidant effect of Green Tea Mega EGCG against electromagnetic radiation-induced oxidative stress in the hippocampus and striatum of rats. *Electromagn. Biol. Med.* **2017**, *36*, 63–73. [CrossRef] [PubMed]
44. Zhou, Z.; Jiang, W.-J.; Wang, Y.-P.; Si, J.-Q.; Zeng, X.-S.; Li, L. CD36-mediated ROS/PI3K/AKT signaling pathway exacerbates cognitive impairment in APP/PS1 mice after noise exposure. *Sci. Total Environ.* **2024**, *952*, 175879. [CrossRef]
45. Zha, L.; Chen, J.; Sun, S.; Mao, L.; Chu, X.; Deng, H.; Cai, J.; Li, X.; Liu, Z.; Cao, W. Soyasaponins can blunt inflammation by inhibiting the reactive oxygen species-mediated activation of PI3K/Akt/NF-kB pathway. *PLoS ONE* **2014**, *9*, e107655. [CrossRef]
46. Zhang, M.; Jang, H.; Nussinov, R. The mechanism of PI3Kα activation at the atomic level. *Chem. Sci.* **2019**, *10*, 3671–3680. [CrossRef]
47. Lv, H.; Wang, J.; Geng, Y.; Xu, T.; Han, F.; Gao, X.J.; Guo, M.Y. Green tea polyphenols inhibit TBBPA-induced lung injury via enhancing antioxidant capacity and modulating the NF-κB pathway in mice. *Food Funct.* **2024**, *15*, 3411–3419. [CrossRef]
48. Wang, X.; Li, Y.; Liu, S.; Yu, X.; Li, L.; Shi, C.; He, W.; Li, J.; Xu, L.; Hu, Z.; et al. Direct activation of RIP3/MLKL-dependent necrosis by herpes simplex virus 1 (HSV-1) protein ICP6 triggers host antiviral defense. *Proc. Natl. Acad. Sci. USA* **2014**, *111*, 15438–15443. [CrossRef]

animals

MDPI

Article

Effects of Low-Protein Amino Acid-Balanced Diets and *Astragalus* Polysaccharides on Production Performance, Antioxidants, Immunity, and Lipid Metabolism in Heat-Stressed Laying Hens

Wenfeng Liu, Xiaoli Wan, Zhiyue Wang and Haiming Yang *

College of Animal Science and Technology, Yangzhou University, Yangzhou 225009, China; dx120220169@stu.yzu.edu.cn (W.L.); wanxl@yzu.edu.cn (X.W.); dkwzy@263.net (Z.W.)
* Correspondence: hmyang@yzu.edu.cn; Tel.: + 86-139-5259-4711

Simple Summary: This study explores how low-protein amino acid-balanced (LPAB) diets with *Astragalus* polysaccharides (APSs) influence laying hens under heat stress. The results show that APSs and LPAB diets can mitigate the adverse effects of heat stress on hens. APSs mainly regulate the secretion of gonadotropin by reducing the concentration of corticosterone in serum, thereby increasing the concentration of follicle-stimulating hormone (FSH) and FSH receptor (FSHR) mRNA expression to regulate primary follicle development and increase the egg production rate. While low-protein diets improve yolk color and boost FSH, they also decrease immunoglobulin and triglyceride levels. Adding APSs enhances immunoglobulin concentrations and helps recover cholesterol levels. LPAB diets and APSs can both enhance cholesterol uptake and suppress its synthesis. Overall, LPAB diets and APSs are deemed valuable feeding strategies for improving production performance and alleviating heat stress in laying hens.

Abstract: The objective of the study was to investigate the effects of low-protein amino acid-balanced (LPAB) diets supplemented with *Astragalus* polysaccharides (APSs) on the production performance, antioxidants, immunity, and biochemical index of laying hens in an elevated-temperature environment. Fifty-two-week-old Hy-Line Brown chickens (n = 768) were randomly divided into four groups, with eight replicates of 24 hens each. The control group was kept at 24 °C with a basal diet (CON), while the treatment groups were exposed to 32 °C and given the following diets: basal (HB), LPAB (HL), and LPAB with 0.5% APSs (HLA). Under heat stress, APSs increased the egg production rate and number of small white follicles, improved the yolk color, and lowered the feed conversion ratio. LPAB diets increased follicle-stimulating hormone, antioxidant enzyme activities, and anti-inflammatory cytokine activity and up-regulated related genes, whereas they reduced stress-related hormones, malondialdehyde concentrations, and triglyceride concentrations and down-regulated related genes. The addition of APSs enhanced immunoglobulin concentrations and cholesterol recovery and altered the expression of related genes. The study found that the adverse effects of high temperatures are directly related to oxidative stress. LAPB diets and APSs relatively alleviate these adverse effects. Therefore, the importance of feeding strategies such as LPAB diets and APSs for laying hens under heat stress conditions has been identified.

Keywords: oxidative stress; follicular development; antioxidant enzymes; immunomodulation; anti-inflammatory cytokine

1. Introduction

Rising ambient temperatures beyond the thermoneutral zone are a primary cause of heat stress in poultry, impairing performance [1], while in southern China, summer heat readily triggers acute or chronic stress in laying hens because they lack sweat glands and have dense plumage [2,3]. Heat stress directly damages proteins and indirectly generates ROS [4]. High temperatures also bring about adverse effects such as reduced feed conversion ratio (FCR), feed intake, body weight, egg production, egg weight, and egg quality, which can lead to severe egg loss [5,6]. In addition, high-temperature environments can lead to an increase in plasma corticosterone (CORT) concentrations and affect humoral and cellular immune responses, potentially leading to immune dysfunction [7,8]. Moreover, a study has shown that heat stress impairs leukocyte protein synthesis, lowers plasma immunoglobulin G (IgG), immunoglobulin M (IgM), and anti-inflammatory cytokine concentrations, and increases the levels of pro-inflammatory cytokines [9].

To counter high temperatures, farmers often use wet curtain cooling and mechanical ventilation. However, these temperature control systems are expensive and hard to maintain consistently in large or open-sided poultry houses. Crude protein has a higher thermic effect because it generates a greater caloric increment [10]. Reducing protein levels in laying hen diets and supplementing with synthetic amino acids to meet essential amino acid (EAA) needs may be a better strategy. Studies have shown that reducing dietary protein levels from 16.5% to 12% under heat stress conditions and supplementing EAAs can improve the resistance of laying hens to heat stress while maintaining laying performance [11]. Dietary supplementation of methionine and L-lysine can improve protein utilization and production performance [12,13].

In addition, *Astragalus* polysaccharides (APSs) play a role due to their anti-inflammatory, immune stimulation, and antioxidant effects [14,15]. In the APS group, the serum levels of tumor necrosis factor-α (TNF-α), interleukin-1β (IL-1β), and interleukin-6 (IL-6) were significantly decreased, and the expression of related genes was also down-regulated [16].

This study investigates the effects of APSs on the production performance, antioxidants, immunity, and lipid metabolism in heat-stressed laying hens based on a low-protein amino acid-balanced diet to provide a reference. This helps to develop a heat stress mitigation strategy that is more suitable for laying hen production systems.

2. Material and Methods

2.1. Experimental Design and Diets

In the same hen house, 768 Hy-Line Brown laying hens (52 weeks old), with nearly the same body weight, were randomly divided into 4 experimental groups, with 8 replicates in each group and 24 hens in each replicate. Laying hens were observed for 10 days to record individual egg production, feed intake, and health status. The experiment was conducted when the hens were between 52 and 58 weeks of age. According to the egg production, individuals with low egg production or seemingly weak egg production were excluded, and then, the remaining hens were randomly divided into groups. The control group (CON) was kept at a constant 24 ± 1 °C, and the treatment groups (TRT) were exposed to 32 ± 1 °C for 7 h per day (heat stress). Experimental laying hens were from Jurong Haoyuan Ecological Agriculture Technology Co., Ltd. (Zhenjiang, China). The control diet based on maize–soybean meal was formulated to supply the nutrient requirements of the birds [17] (Table 1); the crude protein (CP) level in the basal diet was 15.85%, and in the low-protein diet, it was 13.85%. The CON was fed with basal diet, while the TRT groups were fed with basal diet (HB), low-protein amino acid-balanced (LPAB) diet (HL), and LPAB diets supplemented with 0.5% APSs (HLA). During the experimental period, the hens had unrestricted access to feed and water under a lighting programme of 16 h light: 8 h dark. The chicken cage was a three-layer cage with a length of 40 cm × width of 37 cm × height of 37 cm. Each cage was equipped with a trough and two automatic water dispensers. The diets were prepared, respectively, and all the diets used the same batch of corn, soybean meal and other ingredients. The test additive was APSs powder, purity 98% (*w*/*w*), supplied by Xi'an Shennong Technology Co., Ltd. (Xi'an, China). The polysaccharide was extracted from dried *Astragalus membranaceus* roots by combined ethanol–water extraction, followed by deproteinization, decoloration and lyophilization, yielding a water-soluble, off-white powder. Immediately before diet preparation, APSs were first pre-blended with ten times its weight of corn meal to ensure homogeneity, then incorporated into the basal mash at 0.5% (*w*/*w*) using a horizontal ribbon mixer (5 min mixing time). Heat stress was applied in climate-controlled rooms. From 09:00 to 16:00 each day, temperature was maintained at 32 ± 1 °C and relative humidity at 65 ± 5% using a PID-regulated combination of electric heaters and a variable-speed wet-curtain system; outside this period the rooms were held at 24 ± 1 °C.

Table 1. Composition and nutrient levels of basal diets (g/kg air-dry basis).

Ingredients	Basic Diet	Low-Protein Diet
Corn	632	672
Soybean meal	230	163
Wheat bran	33.5	57.5
Calcium dihydrogen phosphate	6.30	6.30
Limestone	91.2	91.7
Sodium chloride	3.00	3.00
DL-Methionine (990 g/kg)	1.50	1.80
L-lysine HCl	0.00	1.60
L-threonine	0.00	0.90
L-tryptophan	0.00	0.40
[1] Premix	2.50	2.50
[2] Nutrient components		
ME (MJ/kg)	11.0	11.0
Crude protein	154	132
Crude fiber	25.9	24.3
Calcium	35.5	35.5
Total phosphorus	5.00	4.90
Nonphytate phosphorus	2.42	2.36
Methionine	3.86	3.88
Lysine	7.81	7.81
Threonine	5.85	5.91
Tryptophan	1.81	1.83

[1] The premix provided following per kg of the diet: VA 225 K IU, VD_3 60 K IU, VE 360 K IU, VK_3 35 mg, VB_{12} 0.30 mg, Nicotinic 700 mg, Cu ($CuSO_4·5H_2O$) 500 mg, Fe ($FeSO_4·H_2O$) 990 mg, Zn ($ZnSO_4·H_2O$) 1130 mg, Mn ($MnSO_4·H_2O$) 1700 mg. [2] The contents of methionine, lysine, threonine and tryptophan were measured values and others were calculated values.

2.2. Animals and Sample Collection

During the experiment, the total egg production, total egg weight, feed intake and mortality number were recorded every day, and the average egg weight, egg production rate, breakage rate (broken egg number/total egg number), mortality rate, average daily feed intake and FCR were counted with repetition as the unit. Three eggs were randomly selected from each replicate on day 40, day 41 and day 42 for egg quality analysis. Eggshell strength was measured by Egg Force Reader (Orka Technology Ltd., West Bountiful, UT, USA). Eggshell thickness was measured by Egg Force Shell Thickness (Orka Technology Ltd., West Bountiful, UT, USA). The egg shape index was calculated as the vertical diameter divided by the horizontal diameter, following the NY/T 823-2020 method [18]. Egg yolk weight and egg weight were measured by electronic balance (precision, 0.01 g) to calculate the relative egg weight. The egg yolk color, albumen height and Haugh unit were measured by Egg Nnalyzer^{TM} (Orka Technology Ltd., West Bountiful, UT, USA). On day 42 of the experimental period, 24 healthy Hy-Line laying hens were randomly selected from each treatment, and blood samples were taken from their brachial veins. The blood samples were centrifuged at $826\times g$ for 10 min to prepare serum, which was then divided into enzyme-free tubes and stored in a -20 °C freezer. Experimental laying hens were euthanized after cervical dislocation, the ovary and liver were removed after death, the ovary was rinsed with cold PBS, and the number of follicles in the ovaries was determined by diameter, small white follicle (SWF, 2–4 mm), big yellow follicle (BYF, 4–10 mm), graded follicle (GF, diameter > 10 mm) and the number of follicles in the ovaries was counted [19], and all visible hierarchical and pre-hierarchical follicles were carefully removed. The remaining ovarian tissue (excluding the follicles) was then rapidly frozen in liquid nitrogen and stored at -80 °C until RNA extraction for determining the mRNA expression of follicle-stimulating hormone receptor (*FSHR*), luteinizing hormone receptor (*LHR*), estrogen receptor 1 (*ESR1*), estrogen receptor 2 (*ESR2*), β-actin, superoxide dismutase (*SOD1*), nuclear factor erythroid 2-related factor 2 (*Nrf2*), glutathione peroxidase (*GPx*), tumor necrosis factor-α (*TNF-α*), IL-1β, interleukin-10 (*IL-10*), interferon-γ (*IFN-γ*), sterol regulatory element-binding protein (*SREBP2*), 3-hydroxy-3-methylglutaryl-CoA (*HMGCR*), low-density lipoprotein receptor (*LDLR*).

2.3. Measurement of Serum Hormone, Antioxidant, Immunoglobulin, Inflammatory and Biochemical Indicators

Serum concentration of adrenocorticotropic hormone (ACTH), CORT, estradiol (E_2), FSH, LH and progesterone (P_4) were measured using a one-step sandwich ELISA with a double antibody kit from Elabscience (Houston, TX, USA). Serum concentration of immunoglobulin A (IgA), IgG, and IgM, inflammatory factors IL-1β, interleukin-6 (IL-6), IL-10, monocyte chemokine-1 (MCP-1), IFN-γ, and tumor necrosis factor-α (TNF-α) were measured using ELISA kits purchased from Shanghai Yubo Biotechnology Co., Ltd. (Shanghai, China). The total protein concentration in the serum was measured using a kit, and the activities of total SOD (T-SOD), GPx, total antioxidant capacity (T-AOC), and malondialdehyde (MDA) were measured using kits purchased from Nanjing Jiancheng Bioengineering Institute (Nanjing, China). Total protein (TP), albumin (ALB), globulin (GLB), serum total cholesterol (T-CHO), high-density lipoprotein cholesterol (HDL-C), LDL cholesterol (LDL-C), triglyceride (TG) and glucose (GLU) content, as well as the activity of aspartate aminotransferase (AST) and alanine aminotransferase (ALT) were measured in the serum using a UniCel DXC 800 Synchron fully automated biochemical analyzer (Beckman Coulter Inc., Brea, CA, USA).

2.4. Expression of Gonadotropin-Releasing Hormone mRNA in the Liver and Ovaries

Total RNA isolation, cDNA synthesis, and quantitative PCR were performed according to the methods described in a previously published paper [20]. The specificity of the primers was confirmed by melting curve analysis. The reference gene was *β-actin*, and the primer sequences for each gene are listed in Table 2. The $2^{-\Delta\Delta Ct}$ method was used to calculate the relative expression levels of the target genes.

Table 2. Sequences of the primers for the target genes.

Target Genes	Primer Sequences (5′−3′)	Size/bp	Accession No.
FSHR	F:ACATTCCCACCAATGCCACA R:AGTGCACCTTATGGACGACG	300	NM_205079.1
LHR	F:GGGCTTTCCCAAGCCTACAT R:TGGTGTCTTTATTGGCGGCT	133	NM_204936.1
ESR1	F:GCTCTCACCCTTCATCCAT R:GACATCCTCTCACGAATGC	150	NC 006090.3
ESR2	F:AGAGAACGCTGTGGGTAT R:TAGGACGACTCACCAACA	187	NC 006092.3
β-actin	F:TGGATGATGATATTGCTGC R:ATCTTCTCCATATCATCCC	253	K02173
SOD1	F:TTGTCTGATGGAGATCATGGCTTC R:TGCTTGCCTTCAGGATTAAAGTGAG	98	NM_205064.1
Nrf2	F:GGGACGGTGACACAGGAACAAC R:TCCACAGCGGGAAATCAGAAAGATC	93	NM_205117.1
GPx	F:ACGGCGCATCTTCCAAAG R:TGTTCCCCCAACCATTTCTC	288	NM_001277853
TNF-α	F:GCCCAGTTCAGATGAGTTGCC R:AAGAGGCCACCACACGACAG	100	NC_052545.1
IL-1β	F:TGCCTGCAGAAGAAGCCTCG R:GACGGGCTCAAAAACCTCCT	204	NC_052553.1
IL-10	F-CAGACCAGCACCAGTCATCA R-TCCCGTTCTCATCCATCTTCTC	163	NM_001004414.2
IFN-γ	F:TGATGGCGTGAAGAAGGTG R:GACTGGCTCCTTTTCCTTTTG	150	AJ012245
SREBP2	F:TGGGCGACATAGACGAGATG R:CACCGCCACCCTGGAAG	102	XM 015289037
HMGCR	F:TTGGATAGAGGGAAGAGGGAAG R:TTGGATAGAGGGAAGAGGGAAG	137	NM 204485.3
LDLR	F:CATCAGCTTCGGGAACCCTC R:CTGTGCACACTCCGCTGT	96	NM 204452.1

2.5. Statistical Analysis

Microsoft Excel 2019 (v. 1808, Microsoft, Redmond, WA, USA) was used for preliminary data processing. Statistical analysis was conducted using SPSS software, version 19.0 (SPSS Inc., Chicago, IL, USA). Prior to all parametric tests, the homogeneity of variances was assessed using Levene's F-test. One-way analysis of variance (ANOVA) was performed to compare differences among the four experimental groups (CON, HB, HL, and HLA), followed by post hoc multiple comparisons using Tukey's test. The experimental design was not a fully factorial arrangement. Specifically, the LPAB and APS treatments were only applied under heat stress conditions (32 °C), while the thermoneutral group (24 °C) received only the basal diet. Therefore, temperature, protein level, and APS supplementation were not treated as independent factors in a multifactorial model. Data were presented as mean and pooled standard error of the mean (SEM), and statistical significance was set at $p < 0.05$.

3. Results

3.1. Production Performance and Follicle Development

As presented in Tables 3 and 4, compared with the CON, the breakage rate, mortality, average daily feed intake and average egg weight of HB, HL and HLA were not significantly different. In terms of egg production rate (EPR), the CON was the highest, and HB and HL were significantly decreased ($p < 0.05$). Compared with CON, HL and HLA, the FCR of HB worsened significantly ($p < 0.05$). The number of SWF in the CON, HL and HLA was more than HB ($p < 0.05$).

Table 3. Effects of *Astragalus* polysaccharides in low-protein diet on production performance of laying hens.

Items [1]	CON	HB	HL	HLA	SEM [2]	*p*-Value
EPR (%)	83.30 ^{a}	73.27 ^{c}	76.22 ^{bc}	78.79 ^{ab}	0.992	<0.001
Breakage rate (%)	0.17	0.34	0.36	0.24	0.040	0.318
Mortality (%)	0	0.15	0.11	0.14	0.046	0.674
ADFI (g)	103.14	106.04	101.70	100.50	0.775	0.059
Average egg weight (g)	64.67	63.22	63.69	63.45	0.848	0.364
FCR (g/g)	1.95 ^{b}	2.30 ^{a}	2.11 ^{b}	2.04 ^{b}	0.035	<0.001

$^{a–c}$ Mean values with different superscripts in the same row indicate significant differences between treatments ($p < 0.05$). $n = 8$. [1] EPR, egg production rate; FCR, feed conversion ratio; ADFI, average daily feed consumption. [2] SEM: standard error of mean.

Table 4. Effects of *Astragalus* polysaccharides in low-protein diet on follicular development of laying hens.

Items [1]	CON	HB	HL	HLA	SEM [2]	*p*-Value
SWF	33.00 ^{a}	26.65 ^{b}	31.00 ^{a}	30.13 ^{a}	0.654	0.001
BYF	6.38	6.88	6.88	7.00	0.265	0.857
GF	5.25	5.00	5.00	5.25	0.188	0.118

^{a,b} Mean values with different superscripts in the same row indicate significant differences between treatments ($p < 0.05$). $n = 6$. [1] SWF, small white follicles; BYF, big yellow follicle; GF, graded follicle. [2] SEM: standard error of mean.

3.2. Egg Quality and Egg Components

According to Table 5, compared to the CON, the LPAB diets deepened yolk color ($p < 0.05$). Temperature, protein level in the diets and APS addition had no significant effect on albumen height, Haugh unit, yolk ratio, eggshell strength, eggshell thickness and egg shape index ($p > 0.05$).

Table 5. Effects of *Astragalus* polysaccharides in low-protein diet on egg quality.

Items [1]	CON	HB	HL	HLA	SEM [2]	*p*-Value
Albumen height (mm)	6.66	6.80	6.68	6.68	0.129	0.312
Haugh unit	78.91	82.40	78.25	84.20	1.051	0.138
Yolk color	5.17 ^{c}	5.85 ^{bc}	6.38 ^{ab}	6.46 ^{a}	0.131	<0.001
Yolk ratio (%)	26.29	25.48	25.43	25.39	0.002	0.115
Eggshell strength (N)	4.17	4.05	4.11	4.19	0.110	0.971
Shell thickness (mm)	0.29	0.29	0.29	0.27	0.005	0.324
Egg shape index	1.31	1.29	1.32	1.30	0.005	0.507

$^{a–c}$ Mean values with different superscripts in the same row indicate significant differences between treatments ($p < 0.05$). $n = 24$. [1] The control group was maintained at 24 °C with a basal diet (CON), while all experimental groups were exposed to 32 °C, fed a basal diet (HB), a low-protein amino acid-balanced (LPAB) diet (HL), LPAB diet with 0.5% *Astragalus* polysaccharides (HLA). [2] SEM: standard error of mean.

3.3. Serum Hormone Indicators and Relevant Gene Expression in the Ovary

According to Table 6, the HB had higher concentration of ACTH, CORT, whereas FSH was lower compared to the other groups ($p < 0.05$). Additionally, the HL had higher concentration of CORT compared to CON and HLA ($p < 0.05$). Figure 1 demonstrated that the *FSHR* in the ovary was up-regulated in CON, HL and HLA ($p < 0.05$).

Table 6. Effects of *Astragalus* polysaccharides on serum hormone indexes in low-protein diet under high-temperature environment.

Items [1]	CON	HB	HL	HLA	SEM [2]	*p*-Value
ACTH (pg/mL)	40.0 [b]	68.3 [a]	39.4 [b]	43.0 [b]	2.36	<0.001
CORT (ng/mL)	69.5 [c]	86.5 [a]	80.3 [b]	73.9 [c]	1.47	<0.001
E_2 (pg/mL)	376	373	366	381	3.45	0.195
FSH (mIU/mL)	12.25 [a]	8.39 [b]	11.70 [a]	12.53 [a]	0.49	0.005
LH (mIU/mL)	14.20	13.71	13.38	13.57	0.56	0.055
P_4 (pmol/mL)	1.20	1.23	1.20	1.27	0.03	0.151

[a–c] Mean values with different superscripts in the same row indicate significant differences between treatments ($p < 0.05$). $n = 6$. [1] The control group was maintained at 24 °C with a basal diet (CON), while all experimental groups were exposed to 32 °C, fed a basal diet (HB), a low-protein amino acid-balanced (LPAB) diet (HL), LPAB diet with 0.5% *Astragalus* polysaccharides (HLA). CORT, corticosterone; E_2, estradiol; FSH, follicle-stimulating hormone; LH, luteinizing hormone; P_4, progesterone. [2] SEM: standard error of mean.

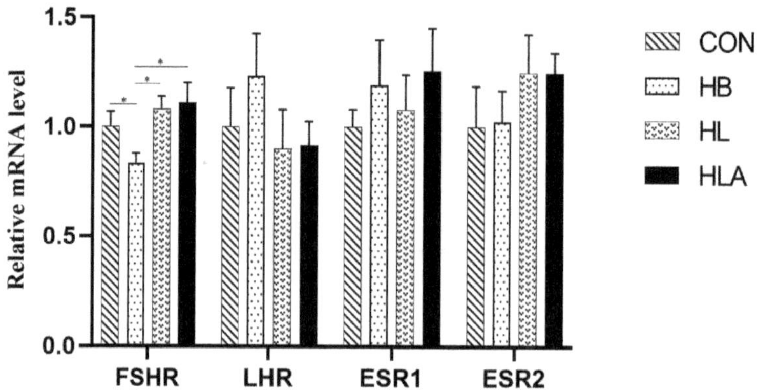

Figure 1. Relative gene expression of productive hormone-related genes. CON: a group was maintained at 24 °C with a basal diet; HB: a group was exposed to 32 °C, fed basal diet; HL: a group was exposed to 32 °C, low-protein amino acid-balanced diet; HLA: a group was exposed to 32 °C low-protein amino acid-balanced diet with 0.5% *Astragalus* polysaccharides. $n = 6$; * indicates $p < 0.05$ relative to the control group. All data are expressed as mean ± SEM in the figure.

3.4. Serum Antioxidant Index and Relevant Gene Expression in the Liver

According to Table 7, T-SOD activity in CON was significantly higher than that in HB and HL groups ($p < 0.05$), and HLA was significantly higher than that in HB group ($p < 0.05$); MDA concentrations in the CON and HLA were lower than those in the HB and HL ($p < 0.05$), and the HLA was higher than the CON ($p < 0.05$). GPx concentration in the CON was higher than in the HB and HL ($p < 0.05$), the HB was lower than all other groups ($p < 0.05$), and the HLA showed no significant difference compared with CON and HL ($p > 0.05$). From Figure 2, the *SOD1* and *GPx* in the liver were up-regulated in CON, HL and HLA compared to HB ($p < 0.05$).

Table 7. Effects of *Astragalus* polysaccharides in low-protein diet on antioxidation performance of serum.

Items [1]	CON	HB	HL	HLA	SEM [2]	p-Value
T-SOD (U/mL)	143.66 [a]	126.51 [c]	133.49 [b]	137.21 [ab]	1.593	<0.001
MDA (nmol/mL)	234.24 [c]	338.78 [a]	329.64 [a]	289.27 [b]	8.231	<0.001
GPx (U/mL)	95.53 [a]	85.48 [c]	88.89 [b]	91.93 [ab]	0.973	<0.001
T-AOC (U/mL)	6.28	5.76	6.41	6.41	0.116	0.145

[a–c] Mean values with different superscripts in the same row indicate significant differences between treatments ($p < 0.05$). $n = 6$. [1] The control group was maintained at 24 °C with a basal diet (CON), while all experimental groups were exposed to 32 °C, fed a basal diet (HB), a low-protein amino acid-balanced (LPAB) diet (HL), LPAB diet with 0.5% *Astragalus* polysaccharides (HLA). T-SOD, total superoxide dismutase; MDA, malondialdehyde; GPx, glutathione peroxidase; T-AOC, total antioxidant capacity. [2] SEM: standard error of mean.

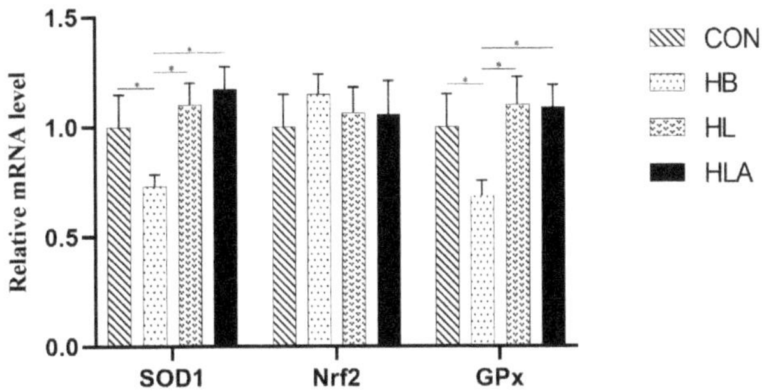

Figure 2. The mRNA expression of antioxidant-related genes. CON: a group was maintained at 24 °C with a basal diet; HB: a group was exposed to 32 °C, fed basal diet; HL: a group was exposed to 32 °C, low-protein amino acid-balanced diet; HLA: a group was exposed to 32 °C low-protein amino acid-balanced diet with 0.5% *Astragalus* polysaccharides. $n = 6$; * indicates $p < 0.05$ relative to the control group. All data are expressed as mean ± SEM in the figure.

3.5. Serum Immunoglobulin, Inflammatory Cytokine Indicators and Relevant Gene Expression in the Liver

As presented in Tables 8 and 9, the HLA had the highest concentrations of IgA, IgG, and IgM compared to the other groups ($p < 0.05$). Specifically, the CON had higher IgA and IgM concentration compared to the HB and HL ($p < 0.05$), while the HB had lower IgG and IgM concentration compared to the HL ($p < 0.05$). IL-1β concentration was higher in the HB compared to other groups ($p < 0.05$), while the HL and HLA were not significantly different from each other ($p > 0.05$), but were both higher than CON ($p < 0.05$). For IL-10, the HLA showed a higher concentration than the other groups ($p < 0.05$), which did not differ significantly from each other ($p > 0.05$). As shown in Figure 3, in the liver, the *IL-1β* mRNA level was highest in HB ($p < 0.01$), and was also higher in HL and HLA than in CON ($p < 0.05$). The relative expression level of *IL-10* mRNA in HLA was significantly higher than that in other groups ($p < 0.05$).

Table 8. Effects of *Astragalus* polysaccharides on serum immunoglobulin content of laying hens under high-temperature environment in low-protein diet.

Items [1]	CON	HB	HL	HLA	SEM [2]	p-Value
IgA (μg/mL)	157.83 [b]	139.11 [c]	144.97 [c]	256.07 [a]	8.69	<0.001
IgG (mg/mL)	2.14 [b]	2.01 [c]	2.11 [b]	2.62 [a]	0.05	<0.001

Table 8. *Cont.*

Items [1]	CON	HB	HL	HLA	SEM [2]	*p*-Value
IgM (μg/mL)	651.80 [b]	482.49 [d]	582.76 [c]	671.91 [a]	13.55	<0.001

[a–d] Mean values with different superscripts in the same row indicate significant differences between treatments ($p < 0.05$). $n = 6$. [1] The control group was maintained at 24 °C with a basal diet (CON), while all experimental groups were exposed to 32 °C, fed a basal diet (HB), a low-protein amino acid-balanced (LPAB) diet (HL), LPAB diet with 0.5% *Astragalus* polysaccharides (HLA). IgA, immunoglobulin A; IgG, immunoglobulin G; IgM, immunoglobulin M. [2] SEM: standard error of mean.

Table 9. Effects of *Astragalus* polysaccharides on serum inflammatory cytokines in laying hens under high-temperature environment in low-protein diet.

Items [1]	CON	HB	HL	HLA	SEM [2]	*p*-Value
IL-1β (pg/mL)	473.35 [c]	598.33 [a]	511.92 [b]	496.64 [b]	8.89	<0.001
IL-2 (pg/mL)	236.64	254.43	256.30	232.06	5.18	0.259
IL-6 (pg/mL)	21.88	26.62	24.61	21.00	0.86	0.074
IL-10 (pg/mL)	48.46 [b]	45.60 [b]	44.45 [b]	61.64 [a]	1.40	<0.001
MCP-1 (ng/mL)	0.58	0.50	0.59	0.53	0.02	0.152
TNF-α (pg/mL)	45.89	53.03	53.35	48.21	6.34	0.127
IFN-γ (pg/mL)	66.00	63.88	69.09	70.84	8.32	0.253

[a–c] Mean values with different superscripts in the same row indicate significant differences between treatments ($p < 0.05$). $n = 6$. [1] The control group was maintained at 24 °C with a basal diet (CON), while all experimental groups were exposed to 32 °C, fed a basal diet (HB), a low-protein amino acid-balanced (LPAB) diet (HL), LPAB diet with 0.5% *Astragalus* polysaccharides (HLA). IL-1β, interleukin 1 beta; IL-2, interleukin 2; IL-6, interleukin 6; IL-10, interleukin 10; MCP-1, monocyte chemokine-1; TNF-α, tumor necrosis factor-α; IFN-γ, interferon-γ. [2] SEM: standard error of mean.

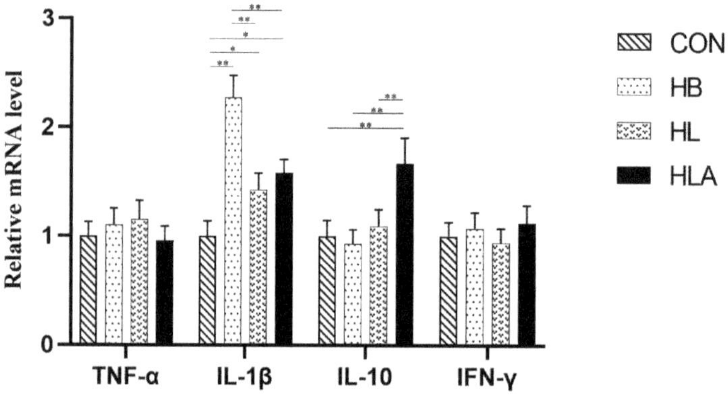

Figure 3. The mRNA expression of inflammatory-related genes. CON: a group was maintained at 24 °C with a basal diet; HB: a group was exposed to 32 °C, fed basal diet; HL: a group was exposed to 32 °C, low-protein amino acid-balanced diet; HLA: a group was exposed to 32 °C low-protein amino acid-balanced diet with 0.5% *Astragalus* polysaccharides. $n = 6$; * indicates $p < 0.05$ and ** indicates $p < 0.01$ relative to the control group. All data are expressed as mean ± SEM in the figure.

3.6. Serum Biochemical Indicators and Relevant Gene Expression in the Liver

According to Table 10, T-CHO concentration was lowest in the CON compared to the other groups ($p < 0.05$). In addition, the HL and HLA had lower T-CHO concentrations and AST activity than the HB ($p < 0.05$). Moreover, the HL exhibited higher concentration of HDL-C than the CON ($p < 0.05$), while the HB showed higher concentration of LDL-C compared to the CON and HLA ($p < 0.05$). Regarding TG and GLU content, the CON had lower concentration than the TRT ($p < 0.05$), while the HL and HLA had lower concentration than the HB ($p < 0.05$). From Figure 4, in the liver, the HMGCR mRNA level was highest in HB ($p < 0.01$), and was also higher in HL and HLA than in CON ($p < 0.05$). LDLR was up-regulated in CON, HL and HLA compared to HB ($p < 0.05$), and also higher in HL and HLA than in CON ($p < 0.05$).

Table 10. Effects of *Astragalus* polysaccharides on serum biochemical indexes under high-temperature environment in low-protein diet.

Items [1]	CON	HB	HL	HLA	SEM [2]	*p*-Value
T-CHO (mmol/L)	2.51 ^{c}	3.91 ^{a}	3.28 ^{b}	3.00 ^{b}	0.29	<0.001
HDL-C (mmol/L)	0.54 ^{b}	0.60 ^{ab}	0.63 ^{a}	0.59 ^{ab}	0.01	0.041
LDL-C (mmol/L)	1.26 ^{b}	1.42 ^{a}	1.33 ^{ab}	1.26 ^{b}	0.02	0.019
TG (mmol/L)	12.63 ^{c}	20.33 ^{a}	16.41 ^{b}	17.96 ^{b}	0.60	<0.001
AST/(U/L)	24.17 ^{a}	24.83 ^{a}	18.83 ^{b}	20.67 ^{b}	0.72	0.004
ALT (U/L)	180.35	186.82	185.01	174.02	5.98	0.313
TP (g/L)	52.50	55.68	54.28	56.17	0.72	0.279
ALB (g/L)	14.39	15.33	15.30	14.99	0.25	0.522
GLB (g/L)	38.00	38.00	39.00	39.33	0.45	0.657
A/G	0.38	0.41	0.39	0.38	0.01	0.791
GLU (mmol/L)	6.04 ^{d}	7.17 ^{c}	7.95 ^{a}	7.67 ^{b}	0.14	<0.001

$^{a–d}$ Mean values with different superscripts in the same row indicate significant differences between treatments ($p < 0.05$). $n = 6$. [1] The control group was maintained at 24 °C with a basal diet (CON), while all experimental groups were exposed to 32 °C, fed a basal diet (HB), a low-protein amino acid-balanced (LPAB) diet (HL), LPAB diet with 0.5% *Astragalus* polysaccharides (HLA). T-CHO, total cholesterol; HDL-C, high-density lipoprotein cholesterol; LDL-C, low-density lipoprotein cholesterol; TG, triglyceride; TP, total protein; ALB, albumin; GLB, globulin; ALT, alanine aminotransferase; AST, aspartate aminotransferase; A/G, albumin/globulin; GLU, glucose. [2] SEM: standard error of mean.

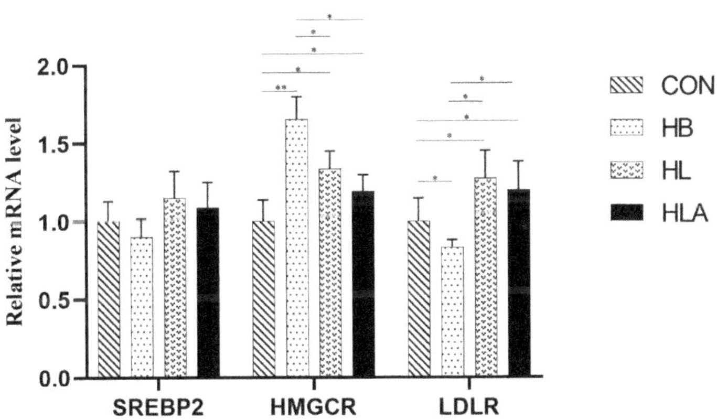

Figure 4. The mRNA expression of reproductive lipid metabolism-related genes. CON: a group was maintained at 24 °C with a basal diet; HB: a group was exposed to 32 °C, fed basal diet; HL: a group was exposed to 32 °C, low-protein amino acid-balanced diet; HLA: a group was exposed to 32 °C low-protein amino acid-balanced diet with 0.5% *Astragalus* polysaccharides. $n = 6$; * indicates $p < 0.05$ and ** indicates $p < 0.01$ relative to the control group. All data are expressed as mean ± SEM in the figure.

4. Discussion

4.1. Production Performance, Follicle Development, Serum Hormone and Related Gene Expression in the Ovary

The results showed that heat stress led to a severe decline in EPR, and that low-protein amino acid-balanced diets and APSs offset some of the effects of heat stress, but still not to the same extent as the CON. Furthermore, the trend in FCR was consistent with EPR. Previous studies have shown that high environmental temperatures disadvantageously influence the production performance of birds, probably due to long-term behavioral, metabolic and physiological changes in response to heat stress [21,22]. A lot of research has shown that high dietary CP can result in reduced protein use and increased heat output, while restricting consumption of protein and adding synthetic DL-methionine and L-lysine could increase the efficiency of hens' protein use and productivity [13,23]. CP has a higher thermal gain and is conducive to more metabolism heat production compared to fat and carbohydrate [24]. Therefore, a low-protein diet effectively mitigates heat stress and boosts protein efficiency in laying hens. This process is achieved by supplementing amino acids (AAs) in order to reduce feeding cost [25,26] and environmental pollution [27]. The mechanism of increased FCR in heat-stressed laying hens fed a low-CP diet is not clear. Reduction in dietary CP (1–3%) did not significantly affect the egg weight and egg production performance of laying hens with a balanced amino acid diet [28,29]. Moreover, it has been reported that if synthetic AA (methionine, lysine, threonine, tryptophan and ILE) is added, the CP in the egg diets can be reduced by 2% [30].

APS has many beneficial properties, such as anti-oxidation [31], immune regulation [32,33], anti-inflammatory [34], hypoglycemic [35] and other biological activities. For increased reactive oxygen species in laying hens due to heat stress, APS can be a kind of inhibiting agent to decrease the side effects of heat stress. Normal ovulation relies on the development of hierarchical follicles [36]. This study showed that the number of SWF in the HL and HLA groups was higher than in the HB group, which may be due to the APSs, which can promote blood circulation of laying hens and further promote follicular development. The LPAB diets can reduce the adverse effects of heat stress, and APSs also have antioxidant and anti-inflammatory effects, which have positive effects on protecting follicular health.

ACTH is the main regulator of adrenal nutrition and fascicular steroidogenesis [37]. Under the conditions of this experiment, the high-temperature environment significantly reduced the ACTH concentration in the serum of laying hens. In birds, gonadotropin-releasing hormone released by the hypothalamus stimulates pituitary secretion of FSH and LH, supporting follicular development and ovulation [38]. FSH and LH can promote the formation of oocytes by stimulating the production of ovarian gonadal hormones (such as E_2, P_4, etc.) [39]. Studies have shown that the release of ACTH and CORT in serum is affected under heat stress [40]. Studies have found that elevated serum ACTH and CORT concentration under heat stress conditions inhibit the secretion of FSH and LH [41]. In this experiment, the addition of APSs in the LPAB diet could reduce the concentration of ACTH and CORT in serum, which was basically consistent with the above results. It can be inferred that the possible mechanism is that the protein is metabolized in the body to provide precursor products for hormone synthesis, or it may be directly involved in the normal regulation of the hypothalamus–pituitary–ovarian axis that affects the secretion of gonadotropin, thereby reducing the synthesis, storage and secretion of FSH and LH, while APSs mainly regulate the secretion of gonadotropin by reducing the concentration of CORT in serum, thereby increasing the concentration of FSH and FSHR mRNA expression to regulate primary follicle development and increase egg production rate.

4.2. *Effects of Astragalus Polysaccharides in Low-Protein Diet on Egg Quality*

Research has shown that most egg quality traits remain unaffected by protein level [42]. Additionally, observations have indicated that eggshell characteristics do not change when the methionine content in the feed is increased [43]. Haugh unit scores were significantly lower in hens fed 13% CP diet than 16% CP diet, but yolk color was significantly higher than in hens fed 16% CP diet [23]. Low-protein diets contain a high proportion of maize, which is rich in lutein; adding more maize to the diet can improve yolk pigmentation [44]. Research indicates that appropriate levels of EAAs can prevent the harmful effects of low-CP diets on shell quality [45]. Eggshell strength was also not influenced by dietary CP levels, suggesting that the addition of L-threonine and lysine to low-crude-protein diets can maintain eggshell quality when digestible sulfur-containing amino acid requirements are considered.

4.3. *Serum Antioxidant and Relevant Gene Expression in the Liver*

The SOD activity and MDA content are important indicators of antioxidant performance. GPx is an important peroxidase enzyme widely present in the body, and is one of the indicators of the body's resistance to peroxidation. It has been shown that high-temperature conditions induce the production of free radicals and oxidative stress, which induce lipid peroxidation and cell membrane damage [46]. A decrease in SOD activity and GPx concentration, consistent with an increase in free radical production in the high-temperature group, has been confirmed [47]. Meanwhile, findings indicate that heat stress increases the levels of SOD and MDA in broiler livers [48]. Nrf-2, a key transcription factor in the antioxidant system, protects cells by upregulating antioxidants like SOD and GPx to counteract stressors [49], which is consistent with the findings of this study. Low-protein diets, supplemented with essential amino acids to meet animals' nutritional needs and maintain amino acid balance, can boost antioxidant enzyme synthesis and enhance antioxidant capacity [50]. Beneficial effects of feeding low-protein diets (with recommended ME) under heat stress have been observed, and this finding is consistent with several other studies conducted under similar conditions [51,52]. Furthermore, it has been shown that oxidative stress dominates the damage caused by heat stress; heat stress causes enhanced metabolism in livestock organisms, causing excessive accumulation of ROS and causing an oxidative–antioxidant imbalance in the organism and oxidative stress; this in turn causes oxidative damage to tissue cells, proteins and nucleic acids [53,54].

4.4. *Serum Immunoglobulin, Inflammatory Cytokines and Relevant Gene Expression in the Liver*

IgA, IgG and IgM play an important role in the immune process of poultry [55]. IgM is an early antibody of humoral immunity, and IgG has the function of neutralizing viruses and bacteria [56]. Supplementation of 0.498% tryptophan in low-protein diet (14.05% CP) significantly increased serum IgG and IgM contents in broilers [57]. Studies have reported that heat stress can inhibit the production of chicken antibodies, and the production of IgA is the lowest [58], which is consistent with the results of [59]. In this experiment, high temperature reduced the content of IgA, IgG and IgM in serum, and low-protein diet significantly increased the content of IgG and IgM, and the low-protein diet supplemented with APSs reversed the adverse effects of high temperature and increased the content of immunoglobulin.

IL-1β is mainly released by monocytes, macrophages and non-immune cells during cell injury and infection, which can promote the production of other cytokines and play an important role in immune response [60]. IL-10 exerts strong anti-inflammatory effects by suppressing pro-inflammatory cytokine production and modulating cytokine receptor expression [61]. Under the conditions of this experiment, the high-temperature environment increased the content of IL-1β in the serum of laying hens. The concentration of IL-1β in the serum of LPAB diets decreased by 14.4% and 17.00%, respectively, and IL-1β relative mRNA expression in the liver was decreased, but it still failed to completely offset the damage caused by heat stress. After adding APSs to the LPAB diet, the concentration of IL-10 increased by 38.67%, and IL-10 relative mRNA expression in the liver was upregulated, indicating that although the high temperature destroyed the balance of its cytokines and adversely affected the immune system of laying hens, the LPAB diets and APSs started the anti-inflammatory reaction. At the same time, it also inhibits inflammation, and reduces body damage.

4.5. Serum Biochemical and Relevant Gene Expression in the Liver

Cholesterol and triglyceride are mainly related to lipid metabolism in laying hens. The contents of CHOL and LDL-C in 21-day-old male broilers were significantly increased under heat stress [62]. APSs could reverse the increase in serum T-CHO, TG and LDL-C levels and the decrease in serum HDL-C levels induced by alcohol in mice [63]. Under the condition of high-temperature stress, the addition of APSs to the diet can significantly reduce the content of total cholesterol and LDL-C in serum, which is basically consistent with the above studies. It may be due to the fact that APSs are good lipid regulators and can improve plasma lipoprotein levels [59,64]. Under the condition of this experiment, heat stress and low-protein diet led to the increase in total cholesterol and triglyceride content in serum of laying hens, which may be due to the decrease in basal metabolism of laying hens in high-temperature environment or low-protein state, affecting the normal metabolism of lipids in vivo, and the decrease in lipid concentration. The energy required for cell metabolism is only provided by related enzymes to decompose nutrients, so the contents of TG and T-CHO involved in lipid metabolism are significantly increased. ALT and AST are liver function markers associated with amino acid metabolism, and their elevated levels under stress reflect tissue damage [65,66]. In this experiment, the addition of APSs in the diet can reduce the activity of AST in serum, protect the liver and reduce the damage caused by heat stress. In different treatments of low-crude-protein diets, no significant changes were observed in TP, ALB, GLB, and AST (standard CP 2, 4, and 6% control diets) [67]. It is speculated that the addition of APSs to the low-protein diet has no obvious effect on improving the protein metabolism of laying hens under high-temperature environment. In addition, heat stress can destroy the integrity of the intestinal barrier, promote the absorption of glucose, and increase serum glucose content [68]. Giving 4% low-protein diet to pregnant mice will lead to dysfunction of islet β cells in their offspring and reduce the secretion of insulin and amylin [69]. Studies have shown that APSs can reduce serum GLU levels, increase insulin sensitivity, and reduce insulin resistance [70]. The results of this experiment are basically consistent with the above results. Some studies found that when broilers were fed low-protein diets (2–4% CP), there was no change in serum total protein [71,72]. Further observations indicated that total protein is only affected when diets lack amino acids [73]. Thus, meeting AA needs seems more critical than CP itself. As shown in this study, low-CP diets can boost hepatic lipogenesis, raising triglyceride levels. Cholesterol biosynthesis starts with acetyl-CoA, with *HMGCR* acting as the rate-limiting enzyme [74]. Cells also take up cholesterol via *LDLR*-mediated endocytosis [75]. Our data showed significantly upregulated *HMGCR* expression in HB livers. Compared to HB, hepatic *LDLR* expression increased by 53% in HL and 45% in HLA, suggesting heat stress boosts blood cholesterol by reducing extrahepatic cholesterol absorption. The LPAB diets and APSs can both enhance cholesterol uptake and suppress its synthesis.

5. Conclusions

In conclusion, heat stress caused by a high-temperature environment reduces production performance. LPAB diets and APSs can reduce the adverse effects of heat stress, improving FCR and follicular development, increasing egg yolk color, and adding 0.5% APSs can also enhance EPR. In addition, LPAB diets and APSs significantly improve antioxidant and immune performance in heat stress. Based on LPAB diets, this study added APSs to explore its effect on laying performance, antioxidant performance, immune performance and lipid metabolism of laying hens. This provides a reference for the use of LPAB diets and plant-derived feed additives in laying hen production. At the same time, it provides a feasible method to mitigate the negative impact of high-temperature environments on laying hen production. APSs can improve the health status of laying hens, and it is expected to be a green, healthy, production-friendly, and antioxidant feed additive.

Author Contributions: Conceptualization, H.Y.; Methodology, H.Y.; Investigation, W.L., Z.W., H.Y.; Data curation, W.L.; Writing—original draft preparation, W.L.; Writing—review and editing, W.L., X.W., Z.W., H.Y.; Supervision, H.Y.; Project administration, H.Y.; Funding acquisition, H.Y. All authors have read and agreed to the published version of the manuscript.

Funding: This research was funded by the Application Demonstration Project of Agricultural Science and Technology Achievements in Jiangsu Province (Rural industry revitalization) (grant number BN2022054); Zhenjiang Key R & D Program-Modern Agriculture (grant number NY2021025). Yangzhou University-Hai'an City Agricultural Science and Technology Modernization Pilot County Project (grant number HA-XXY-2023-01).

Institutional Review Board Statement: All animal experiments were conducted in accordance with the Institutional Animal Care and Use Committee of Yang Zhou University, Yang Zhou, China (Approval No. SYXK(Su)2022-0020, approved on 15 March 2022).

Informed Consent Statement: Not applicable.

Data Availability Statement: The original contributions presented in this study are included in the article. Further inquiries can be directed to the corresponding author.

Acknowledgments: The authors would like to thank Jiasen Jiang for technical support and the barn staff for managing the animals.

Conflicts of Interest: The authors declare no conflicts of interest.

References

1. Tučková, K.; Filipčík, R. Effect of Heat Stress on Age in Conception of Holstein Heifers. *Acta. Univ. Agric. Silvic. Mendel. Brun.* **2019**, *67*, 173–177. [CrossRef]
2. Barrett, N.W.; Rowland, K.; Schmidt, C.J.; Lamont, S.J.; Rothschild, M.F.; Ashwell, C.M.; Persia, M.E. Effects of Acute and Chronic Heat Stress on the Performance, Egg Quality, Body Temperature, and Blood Gas Parameters of Laying Hens. *Poult. Sci.* **2019**, *98*, 6684–6692. [CrossRef]
3. Fatima, N.; Ahmad, M.; Usman, M.; Haider, U.; Raza Farhan, M.H.; Iftikhar, E.; Hassan, S.U.; Narayan, E. Role of Feed Additives in Mitigating the Impact of Heat Stress on Poultry Gut Health and Productivity. *World's Poult. Sci. J.* **2024**, *80*, 1055–1075. [CrossRef]
4. Vandana, G.D.; Sejian, V.; Lees, A.M.; Pragna, P.; Silpa, M.V.; Maloney, S.K. Heat Stress and Poultry Production: Impact and Amelioration. *Int. J. Biometeorol.* **2021**, *65*, 163–179. [CrossRef] [PubMed]
5. Ezzat, W.; Mahrose, K.M.; Rizk, A.M.; Ouda, M.M.M.; Fathey, I.A.; Othman, S.I.; Allam, A.A.; Rudayni, H.A.; Almasmoum, H.A.; Taha, A.E.; et al. Impact of β-Glucan Dietary Supplementation on Productive, Reproductive Performance and Physiological Response of Laying Hens under Heat Stress Conditions. *Poult. Sci.* **2024**, *103*, 103183. [CrossRef] [PubMed]
6. Saleh, A.A.; Eltantawy, M.S.; Gawish, E.M.; Younis, H.H.; Amber, K.A.; Abd El-Moneim, A.E.-M.E.; Ebeid, T.A. Impact of Dietary Organic Mineral Supplementation on Reproductive Performance, Egg Quality Characteristics, Lipid Oxidation, Ovarian Follicular

Development, and Immune Response in Laying Hens Under High Ambient Temperature. *Biol. Trace Elem. Res.* **2020**, *195*, 506–514. [CrossRef] [PubMed]
7. Johnson, J.S.; Maskal, J.M.; Duttlinger, A.W.; Kpodo, K.R.; McConn, B.R.; Byrd, C.J.; Richert, B.T.; Marchant-Forde, J.N.; Lay, D.C.; Perry, S.D.; et al. In Utero Heat Stress Alters the Postnatal Innate Immune Response of Pigs. *J. Anim. Sci.* **2020**, *98*, skaa356. [CrossRef]
8. Siddiqui, S.H.; Kang, D.; Park, J.; Khan, M.; Belal, S.A.; Shin, D.; Shim, K. Altered Relationship between Gluconeogenesis and Immunity in Broilers Exposed to Heat Stress for Different Durations. *Poult. Sci.* **2021**, *100*, 101274. [CrossRef]
9. Benyelloul, K.; Seddik, L.; Bouhadda, Y.; Bououdina, M.; Aourag, H.; Khodja, K. Effect of Pressure on Structural, Elastic and Mechanical Properties of Transition Metal Hydrides Mg7TMH16 (TM = Sc, Ti, V, Y, Zr and Nb): First-Principles Investigation. *J. Phys. Chem. Solids* **2017**, *111*, 229–237. [CrossRef]
10. Teyssier, J.-R.; Brugaletta, G.; Sirri, F.; Dridi, S.; Rochell, S.J. A Review of Heat Stress in Chickens. Part II: Insights into Protein and Energy Utilization and Feeding. *Front. Physiol.* **2022**, *13*, 943612. [CrossRef]
11. Torki, M.; Mohebbifar, A.; Ghasemi, H.A.; Zardast, A. Response of Laying Hens to Feeding Low-Protein Amino Acid-Supplemented Diets under High Ambient Temperature: Performance, Egg Quality, Leukocyte Profile, Blood Lipids, and Excreta pH. *Int. J. Biometeorol.* **2015**, *59*, 575–584. [CrossRef]
12. Liu, Y.; Wang, D.; Zhao, L.; Zhang, J.; Huang, S.; Ma, Q. Effect of Methionine Deficiency on the Growth Performance, Serum Amino Acids Concentrations, Gut Microbiota and Subsequent Laying Performance of Layer Chicks. *Front. Vet. Sci.* **2022**, *9*, 878107. [CrossRef] [PubMed]
13. Poosuwan, K.; Bunchasak, C.; Kaewtapee, C. Long-Term Feeding Effects of Dietary Protein Levels on Egg Production, Immunocompetence and Plasma Amino Acids of Laying Hens in Subtropical Condition. *J. Anim. Physiol. Anim. Nutr.* **2010**, *94*, 186–195. [CrossRef] [PubMed]
14. Wang, X.-Y.; Wang, R.-C.; Qu, Z.-Y.; Zhu, Y.-Z.; Li, Y.-L. Advances on Immunoregulation Effect of *Astragalus* Polysaccharides. *Front. Nat. Prod.* **2022**, *1*, 971679. [CrossRef]
15. Lu, Y.; Wu, Y.; Sun, L.; Yang, S.; Kuang, H.; Li, R.; Meng, Y.; Wu, Y. Identifying the Anti-Inflammatory Effects of *Astragalus* Polysaccharides in Anti-N-Methyl-D-Aspartate Receptor Encephalitis: Network Pharmacology and Experimental Validation. *Comb. Chem. High Throughput Screen.* **2024**, *27*, 1022–1032. [CrossRef]
16. Qiao, Y.; Liu, C.; Guo, Y.; Zhang, W.; Guo, W.; Oleksandr, K.; Wang, Z. Polysaccharides Derived from *Astragalus membranaceus* and *Glycyrrhiza uralensis* Improve Growth Performance of Broilers by Enhancing Intestinal Health and Modulating Gut Microbiota. *Poult. Sci.* **2022**, *101*, 101905. [CrossRef]
17. National Research Council, and Subcommittee on Poultry Nutrition. *Nutrient Requirements of Poultry: 1994*; National Academies Press: Washington, WA, USA, 1994; ISBN 0-309-04892-3.
18. *NY/T 823-2020*; Terminology and Measurement Calculation Method for Poultry Performance. The National Standards of the People's Republic of China: Beijing, China, 2020.
19. Palmer, S.S.; Bahr, J.M. Follicle Stimulating Hormone Increases Serum Oestradial-17ß Concentrations, Number of Growing Follicles and Yolk Deposition in Aging Hens (Gallus Gallus Domesticus) with Decreased Egg Production. *Br. Poult. Sci.* **1992**, *33*, 403–414. [CrossRef]
20. Osman, R.H.; Liu, L.; Xia, L.; Zhao, X.; Wang, Q.; Sun, X.; Zhang, Y.; Yang, B.; Zheng, Y.; Gong, D. Fads1 and 2 Are Promoted to Meet Instant Need for Long-Chain Polyunsaturated Fatty Acids in Goose Fatty Liver. *Mol. Cell. Biochem.* **2016**, *418*, 103–117. [CrossRef]
21. Bohler, M.W.; Chowdhury, V.S.; Cline, M.A.; Gilbert, E.R. Heat Stress Responses in Birds: A Review of the Neural Components. *Biology* **2021**, *10*, 1095. [CrossRef]
22. Kim, H.-R.; Ryu, C.; Lee, S.-D.; Cho, J.-H.; Kang, H. Effects of Heat Stress on the Laying Performance, Egg Quality, and Physiological Response of Laying Hens. *Animals* **2024**, *14*, 1076. [CrossRef]
23. Torki, M.; Nasiroleslami, M.; Ghasemi, H.A. The Effects of Different Protein Levels in Laying Hens under Hot Summer Conditions. *Anim. Prod. Sci.* **2016**, *57*, 927–934. [CrossRef]
24. Soares, K.R.; Lara, L.J.C.; da Silva Martins, N.R.; e Silva, R.R.; Pereira, L.F.P.; Cardeal, P.C.; Teixeira, M.d.P.F. Protein Diets for Growing Broilers Created under a Thermoneutral Environment or Heat Stress. *Anim. Feed. Sci. Technol.* **2020**, *259*, 114332. [CrossRef]
25. Mosca, F.; Kuster, C.A.; Stella, S.; Farina, G.; Madeddu, M.; Zaniboni, L.; Cerolini, S. Growth Performance, Carcass Characteristics and Meat Composition of Milanino Chickens Fed on Diets with Different Protein Concentrations. *Br. Poult. Sci.* **2016**, *57*, 531–537. [CrossRef] [PubMed]
26. Zhang, B.; Zhou, S.; Zhai, W.; Zhao, Y. Effect of Reduction in Dietary Amino Acids and Energy on Growth Performance and Economic Return of Cobb 700 and Ross 708 Broilers. *Animals* **2025**, *15*, 890. [CrossRef] [PubMed]

27. Niu, J.; Qiao, Y.; Yang, X.; Chen, X.; Li, H.; Guo, Y.; Zhang, W.; Wang, Z. Protease and Bacillus Coagulans Supplementation in a Low-Protein Diet Improves Broiler Growth, Promotes Amino Acid Transport Gene Activity, Strengthens Intestinal Barriers, and Alters the Cecal Microbial Composition. *Animals* **2025**, *15*, 170. [CrossRef] [PubMed]
28. Liu, H.; Xu, K.; Wang, H.; Lin, H.; Yang, X.; Wang, X.; Zhao, J.; Ma, B.; Shu, Q.; Lu, Y. Effects of Different Forms of Amino Acid Supplementation on the Performance and Intestinal Barrier Function of Laying Hens Fed a Low-Protein Diet. *Poult. Sci.* **2024**, *103*, 104375. [CrossRef]
29. Cabezas-Garcia, E.H.; Rodríguez-Aguilar, D.E.; Afanador-Téllez, G. Individual Egg Production of Hy-Line Brown Hens during the Early Laying Phase in Response to Dietary CP Levels. *Anim.-Open Space* **2022**, *1*, 100027. [CrossRef]
30. Parenteau, I.A.; Stevenson, M.; Kiarie, E.G. Egg Production and Quality Responses to Increasing Isoleucine Supplementation in Shaver White Hens Fed a Low Crude Protein Corn-Soybean Meal Diet Fortified with Synthetic Amino Acids between 20 and 46 Weeks of Age. *Poult. Sci.* **2020**, *99*, 1444–1453. [CrossRef]
31. Wang, J.-M.; Sun, X.-Y.; Ouyang, J.-M. Structural Characterization, Antioxidant Activity, and Biomedical Application of *Astragalus* Polysaccharide Degradation Products. *Int. J. Polym. Sci.* **2018**, *2018*, 5136185. [CrossRef]
32. Liao, J.; Li, C.; Huang, J.; Liu, W.; Chen, H.; Liao, S.; Chen, H.; Rui, W. Structure Characterization of Honey-Processed *Astragalus* Polysaccharides and Its Anti-Inflammatory Activity in Vitro. *Molecules* **2018**, *23*, 168. [CrossRef]
33. Li, K.; Cao, Y.; Jiao, S.; Du, G.; Du, Y.; Qin, X. Structural Characterization and Immune Activity Screening of Polysaccharides with Different Molecular Weights from Astragali Radix. *Front. Pharmacol.* **2020**, *11*, 582091. [CrossRef] [PubMed]
34. Wu, J.; Li, C.; Bai, L.; Wu, J.; Bo, R.; Ye, M.; Huang, L.; Chen, H.; Rui, W. Structural Differences of Polysaccharides from *Astragalus* before and after Honey Processing and Their Effects on Colitis Mice. *Int. J. Biol. Macromol.* **2021**, *182*, 815–824. [CrossRef]
35. Zexi, Z.; Lei, Z.; Hansong, X. Effect of *Astragalus* Polysaccharide in Treatment of Diabetes Mellitus: A Narrative Review. *J. Tradit. Chin. Med.* **2019**, *39*, 133–138. [CrossRef]
36. Yang, L.; Fan, X.; Tian, K.; Yan, S.; Xu, C.; Tian, Y.; Xiao, C.; Jia, X.; Shi, J.; Bai, Y.; et al. Dynamic Expression Profile of Follicles at Different Stages in High- and Low-Production Laying Hens. *Genes* **2023**, *15*, 40. [CrossRef]
37. Tuğalay, Ç.Ş.; Bayraktar, Ö.H.; Genç, N. Effects of ACTH and Acute Heat Stress on Oxidative Stress in an Early Environmentally Enriched Broilers. *J. Anim. Prod.* **2021**, *62*, 93–98.
38. Mishra, S.K.; Chen, B.; Zhu, Q.; Xu, Z.; Ning, C.; Yin, H.; Wang, Y.; Zhao, X.; Fan, X.; Yang, M.; et al. Transcriptome Analysis Reveals Differentially Expressed Genes Associated with High Rates of Egg Production in Chicken Hypothalamic-Pituitary-Ovarian Axis. *Sci. Rep.* **2020**, *10*, 5976. [CrossRef]
39. Yin, Z.Z.; Dong, X.Y.; Cao, H.Y.; Mao, H.G.; Ma, Y.Z. Effects of Rearing Systems on Reproductive Hormones Secretion and Their Receptors Gene Expression in Xianju Chickens under Summer Conditions. *Poult. Sci.* **2018**, *97*, 3092–3096. [CrossRef]
40. Titto, C.G.; Negrão, J.A.; Canaes, T.D.S.; Titto, R.M.; Leme-dos Santos, T.M.D.C.; Henrique, F.L.; Calviello, R.F.; Pereira, A.M.F.; Titto, E.A.L. Heat Stress and ACTH Administration on Cortisol and Insulin-like Growth Factor I (IGF-I) Levels in Lactating Holstein Cows. *J. Appl. Anim. Res.* **2017**, *45*, 1–7. [CrossRef]
41. Huang, Y.; Cai, H.; Han, Y.; Yang, P. Mechanisms of Heat Stress on Neuroendocrine and Organ Damage and Nutritional Measures of Prevention and Treatment in Poultry. *Biology* **2024**, *13*, 926. [CrossRef]
42. Alagawany, M.; El-Hindawy, M.M.; El-Hack, M.E.A.; Arif, M.; El-Sayed, S.A. Influence of Low-Protein Diet with Different Levels of Amino Acids on Laying Hen Performance, Quality and Egg Composition. *An. Acad. Bras. Ciências* **2020**, *92*, e20180230. [CrossRef] [PubMed]
43. van Harn, J.; Dijkslag, M.A.; van Krimpen, M.M. Effect of Low Protein Diets Supplemented with Free Amino Acids on Growth Performance, Slaughter Yield, Litter Quality, and Footpad Lesions of Male Broilers. *Poult. Sci.* **2019**, *98*, 4868–4877. [CrossRef] [PubMed]
44. Zurak, D.; Slovenec, P.; Janječić, Z.; Bedeković, X.; Pintar, J.; Kljak, K. Overview on Recent Findings of Nutritional and Non-Nutritional Factors Affecting Egg Yolk Pigmentation. *World's Poult. Sci. J.* **2022**, *78*, 531–560. [CrossRef]
45. Zhang, H.; Xuan, Y.; Guo, D.; Zeng, Q.; Bai, S.; Liu, Y.; Ding, X.; Zhang, K.; Wang, J. Effects of Dietary Low Protein Levels and Amino Acid Patterns on Production Performance, Egg Quality and Intestinal Function in Laying Hens. *Poult. Sci.* **2025**, *104*, 105578. [CrossRef]
46. Akbarian, A.; Michiels, J.; Degroote, J.; Majdeddin, M.; Golian, A.; De Smet, S. Association between Heat Stress and Oxidative Stress in Poultry; Mitochondrial Dysfunction and Dietary Interventions with Phytochemicals. *J. Anim. Sci. Biotechnol.* **2016**, *7*, 37. [CrossRef]
47. Cheng, Y.; Du, M.; Xu, Q.; Chen, Y.; Wen, C.; Zhou, Y. Dietary Mannan Oligosaccharide Improves Growth Performance, Muscle Oxidative Status, and Meat Quality in Broilers under Cyclic Heat Stress. *J. Therm. Biol* **2018**, *75*, 106–111. [CrossRef]
48. Tang, L.-P.; Liu, Y.-L.; Zhang, J.-X.; Ding, K.-N.; Lu, M.-H.; He, Y.-M. Heat Stress in Broilers of Liver Injury Effects of Heat Stress on Oxidative Stress and Autophagy in Liver of Broilers. *Poult. Sci.* **2022**, *101*, 102085. [CrossRef]

49. Mahanty, A.; Mohanty, S.; Mohanty, B.P. Dietary Supplementation of Curcumin Augments Heat Stress Tolerance through Upregulation of *Nrf-2*-Mediated Antioxidative Enzymes and *Hsps* in *Puntius Sophore*. *Fish Physiol. Biochem.* **2017**, *43*, 1131–1141. [CrossRef]
50. Ouyang, J.; Li, Q.; Zhou, H.; Li, G.; Wu, Y.; Yang, L. Effect of Dietary Tryptophan Supplementation on Serum Biochemical Indices, Antioxidant Capacities, Cytokine Levels and Mitochondrial Function of Broilers under Chronic Heat Stress. *Trop. Anim. Health Prod.* **2023**. [CrossRef]
51. Zulkifli, I.; Akmal, A.F.; Soleimani, A.F.; Hossain, M.A.; Awad, E.A. Effects of Low-Protein Diets on Acute Phase Proteins and Heat Shock Protein 70 Responses, and Growth Performance in Broiler Chickens under Heat Stress Condition. *Poult. Sci.* **2018**, *97*, 1306–1314. [CrossRef]
52. Salahi, A.; Shahir, M.H.; Attia, Y.A.; Fahmy, K.N.E.; Bovera, F.; Tufarelli, V. Impact of Low-Protein Diets on Broiler Nutrition, Production Sustainability, Gene Expression, Meat Quality and Greenhouse Gas Emissions. *J. Appl. Anim. Res.* **2025**, *53*, 2473419. [CrossRef]
53. Belhadj Slimen, I.; Najar, T.; Ghram, A.; Abdrrabba, M. Heat Stress Effects on Livestock: Molecular, Cellular and Metabolic Aspects, a Review. *J. Anim. Physiol. Anim. Nutr.* **2016**, *100*, 401–412. [CrossRef]
54. Surai, P.F.; Kochish, I.I.; Fisinin, V.I.; Kidd, M.T. Antioxidant Defence Systems and Oxidative Stress in Poultry Biology: An Update. *Antioxidants* **2019**, *8*, 235. [CrossRef]
55. Liao, X.D.; Ma, G.; Cai, J.; Fu, Y.; Yan, X.Y.; Wei, X.B.; Zhang, R.J. Effects ofClostridium Butyricum on Growth Performance, Antioxidation, and Immune Function of Broilers. *Poult. Sci.* **2015**, *94*, 662–667. [CrossRef]
56. Song, X.; Qian, J.; Wang, C.; Wang, D.; Zhou, J.; Zhao, Y.; Wang, W.; Li, J.; Guo, R.; Li, Y. Correlation between the IgG/IgA Antibody Response against PEDV Structural Protein and Virus Neutralization. *Microbiol. Spectr.* **2023**, *11*, e05233-22. [CrossRef]
57. Zhou, J.; Qiu, K.; Wang, J.; Zhang, H.; Qi, G.; Wu, S. Effect of Dietary Serine Supplementation on Performance, Egg Quality, Serum Indices, and Ileal Mucosal Immunity in Laying Hens Fed a Low Crude Protein Diet. *Poult. Sci.* **2021**, *100*, 101465. [CrossRef]
58. Sun, H.; Zhao, L.; Xu, Z.-J.; De Marco, M.; Briens, M.; Yan, X.-H.; Sun, L.-H. Hydroxy-Selenomethionine Improves the Selenium Status and Helps to Maintain Broiler Performances under a High Stocking Density and Heat Stress Conditions through a Better Redox and Immune Response. *Antioxidants* **2021**, *10*, 1542. [CrossRef]
59. Alagawany, M.; Ashour, E.A.; El-Fakhrany, H.H.H.; Ismail, T.A.; Nasr, M. Early Nutrition Programming with *Astragalus membranaceus* Polysaccharide: Its Effect on Growth, Carcasses, Immunity, Antioxidants, Lipid Profile and Liver and Kidney Functions in Broiler Chickens. *Anim. Biotechnol.* **2022**, *33*, 362–368. [CrossRef]
60. Kim, D.H.; Lee, W.-W. IL-1 Receptor Dynamics in Immune Cells: Orchestrating Immune Precision and Balance. *Immune Netw.* **2024**, *24*, e21. [CrossRef]
61. Martinez-Espinosa, I.; Serrato, J.A.; Ortiz-Quintero, B. Role of IL-10-Producing Natural Killer Cells in the Regulatory Mechanisms of Inflammation during Systemic Infection. *Biomolecules* **2021**, *12*, 4. [CrossRef]
62. Bueno, J.P.R.; de Mattos Nascimento, M.R.B.; da Silva Martins, J.M.; Marchini, C.F.P.; Gotardo, L.R.M.; de Sousa, G.M.R.; Mundim, A.V.; Guimarães, E.C.; Rinaldi, F.P. Effect of Age and Cyclical Heat Stress on the Serum Biochemical Profile of Broiler Chickens. *Semin. Ciências Agrárias* **2017**, *38*, 1383. [CrossRef]
63. Zhou, J.; Zhang, N.; Zhao, L.; Wu, W.; Zhang, L.; Zhou, F.; Li, J. *Astragalus* Polysaccharides and Saponins Alleviate Liver Injury and Regulate Gut Microbiota in Alcohol Liver Disease Mice. *Foods* **2021**, *10*, 2688. [CrossRef] [PubMed]
64. Saeed, M.; YaTao, X.; Rehman, Z.U.; Arain, M.A.; Soomro, R.N.; El-Hack, M.E.A.; Bhutto, Z.A.; Abbasi, B.; Dhama, K.; Sarwar, M.; et al. Nutritional and Healthical Aspects of Yacon (*Smallanthus sonchifolius*) for Human, Animals and Poultry. *Int. J. Pharmacol.* **2017**, *13*, 361–369. [CrossRef]
65. Mokondjimobe, E.; Longo-Mbenza, B.; Akiana, J.; Ndalla, U.O.; Dossou-Yovo, R.; Mboussa, J.; Parra, H.-J. Biomarkers of Oxidative Stress and Personalized Treatment of Pulmonary Tuberculosis: Emerging Role of Gamma-Glutamyltransferase. *Adv. Pharmacol. Sci.* **2012**, *2012*, 465634. [CrossRef]
66. Ahmed, O.M.; Elkomy, M.H.; Fahim, H.I.; Ashour, M.B.; Naguib, I.A.; Alghamdi, B.S.; Mahmoud, H.U.R.; Ahmed, N.A. Rutin and Quercetin Counter Doxorubicin-induced Liver Toxicity in Wistar Rats via Their Modulatory Effects on Inflammation, Oxidative Stress, Apoptosis, and Nrf2. *Oxidative Med. Cell. Longev.* **2022**, *2022*, 2710607. [CrossRef]
67. Badawi, M.E.-S.; Ali, A.H.; El-Razik, W.M.A.; Soliman, M.H. Influence of Low Crude Protein Diets on Broiler Chickens Performance. *Adv. Anim. Vet. Sci.* **2019**, *7*, 26–33. [CrossRef]
68. Sun, X.; Zhang, H.; Sheikhahmadi, A.; Wang, Y.; Jiao, H.; Lin, H.; Song, Z. Effects of Heat Stress on the Gene Expression of Nutrient Transporters in the Jejunum of Broiler Chickens (*Gallus gallus domesticus*). *Int. J. Biometeorol.* **2015**, *59*, 127–135. [CrossRef]
69. Mathias, P.C.d.F.; Miranda, G.D.S.; Barella, L.F.; Miranda, R.A.; Pavanello, A.; Martins, I.P.; Facchi, J.C.; Costermani, H.d.O.; de Lima, T.A.L.; de Oliveira, J.C. Cholinergic-Pathway-Weakness-Associated Pancreatic Islet Dysfunction: A Low-Protein-Diet Imprint Effect on Weaned Rat Offspring. *J. Dev. Origins Health Dis.* **2020**, *11*, 484–491. [CrossRef]

70. Zheng, Y.; Ren, W.; Zhang, L.; Zhang, Y.; Liu, D.; Liu, Y. A Review of the Pharmacological Action of *Astragalus* Polysaccharide. *Front. Pharmacol.* **2020**, *11*, 349. [CrossRef] [PubMed]
71. Hernández, F.; López, M.; Martínez, S.; Megías, M.D.; Catalá, P.; Madrid, J. Effect of Low-Protein Diets and Single Sex on Production Performance, Plasma Metabolites, Digestibility, and Nitrogen Excretion in 1-to 48-Day-Old Broilers. *Poult. Sci.* **2012**, *91*, 683–692. [CrossRef]
72. Saleh, A.A.; Amber, K.A.; Soliman, M.M.; Soliman, M.Y.; Morsy, W.A.; Shukry, M.; Alzawqari, M.H. Effect of Low Protein Diets with Amino Acids Supplementation on Growth Performance, Carcass Traits, Blood Parameters and Muscle Amino Acids Profile in Broiler Chickens under High Ambient Temperature. *Agriculture* **2021**, *11*, 185. [CrossRef]
73. Ahmadi, M.; Yaghobfar, A.; Tabatabaei, S.H. Study of Effects Difference Levels of Crude Protein and Amino Acid of Diet on Intestinal Morphological and Blood Biological Parameters of Poultry. *Biological Forum* **2015**, *7*, 666–670.
74. Luo, J.; Yang, H.; Song, B.-L. Mechanisms and Regulation of Cholesterol Homeostasis. *Nat. Rev. Mol. Cell Biol.* **2020**, *21*, 225–245. [CrossRef] [PubMed]
75. Siddiqui, H.; Yevstigneyev, N.; Madani, G.; McCormick, S. Approaches to Visualising Endocytosis of LDL-Related Lipoproteins. *Biomolecules* **2022**, *12*, 158. [CrossRef] [PubMed]

animals

MDPI

Article

Effects of Dietary Baicalin on Growth Performance, Serum Biochemical Parameters, Liver Health, Intestinal Health, and Microbiota of Yellow Catfish (*Pelteobagrus fulvidraco*)

Haonan Liu [1,2], Xinru Li [1,2], Yang Fan [1,2], Yang Xiao [1,2], Yunfeng Chen [1,2], Xiaoqin Li [1,2,*] and Xiangjun Leng [1,2,*]

1 National Demonstration Center for Experimental Fisheries Science Education, Shanghai Ocean University, Shanghai 201306, China; haonanliu2002@163.com (H.L.); m15237303262@163.com (X.L.); 19850226601@163.com (Y.F.); xy2000224@163.com (Y.X.); chenyf256893@163.com (Y.C.)

2 Research Centre of the Ministry of Agriculture and Rural Affairs on Environmental Ecology and Fish Nutrition, Shanghai Ocean University, Shanghai 201306, China

* Correspondence: xqli@shou.edu.cn (X.L.); xjleng@shou.edu.cn (X.L.)

Simple Summary: Baicalin is the major bioactive compound in the Chinese herb *Scutellaria baicalensis*, and its positive effects have been reported in some fish. The present study investigated the effects of dietary baicalin on the growth and health of yellow catfish (*Pelteobagrus fulvidraco*). The results indicated that dietary baicalin supplementation significantly improved the growth performance, antioxidant capacity, and morphology of the liver and intestine, as well as positively modulated the intestinal microbiota composition of yellow catfish. The optimal inclusion of dietary baicalin was 400 mg/kg. This finding will guide the application of baicalin in aquatic feeds.

Abstract: This study investigated the effects of dietary baicalin on the growth and health of yellow catfish (*Pelteobagrus fulvidraco*). Five diets with the addition of 0 (control), 100, 200, 400, and 800 mg/kg of baicalin were fed to juvenile yellow catfish (11.19 $\pm$ 0.07 g) for 56 days. Dietary supplementation with 200 or 400 mg/kg baicalin increased weight gain ($p < 0.05$) and reduced the feed conversion ratio (FCR) ($p < 0.05$). The BA400 group exhibited higher total antioxidant capacity, catalase, superoxide dismutase activity, and lower malondialdehyde level in the liver and intestine than the control ($p < 0.05$). In the 400 mg/kg baicalin group, the hepatic vacuolization rate was lower ($p < 0.05$), and the intestinal villus width and villus height were higher than those in the control ($p < 0.05$). In terms of intestinal microbiota composition, the 400 mg/kg baicalin group demonstrated lower abundances of *Proteobacteria* and *Actinobacteria* (phylum level) and *Cetobacterium*, *Candidatus_Arthromitus*, and *Plesiomonas* (genus level) than the control ($p < 0.05$), while the relative abundances of *Fusobacteriota* and *Lactobacillus* (ZOR0006) increased in this group ($p < 0.05$). In conclusion, dietary baicalin supplementation significantly improved the growth, antioxidant capacity, and intestinal morphology and positively modulated the intestinal microbiota composition of yellow catfish. The optimal dietary supplementation level of baicalin is 400 mg/kg for yellow catfish.

Keywords: baicalin; *Pelteobagrus fulvidraco*; growth performance; antioxidant levels; liver and intestinal health; intestinal microbiota

1. Introduction

Baicalin ($C_{21}H_{18}O_{11}$), a yellow bitter flavonoid extracted from the roots of *Scutellaria baicalensis*, is a key bioactive compound in this Chinese medicinal herb [1]. It exhibits various pharmacological effects, including antioxidant, anti-inflammatory, immune-enhancing, antibacterial, hepatoprotective, and intestinal protective properties [2]. Baicalin has been shown to improve the growth performance, antioxidant capacity, and immune function in land animals such as preweaned calves [3], broilers [4], ducks [5], and piglets [6]. In aquatic animals, baicalin was reported to enhance the growth performance, antioxidant capacity, and resistance to hepatic oxidative stress and *Streptococcus agalactiae* infection in Nile tilapia (*Oreochromis niloticus*) [7,8]. Baicalin has also been shown to protect giant river prawns (*Macrobrachium rosenbergii*) against *Vibrio parahaemolyticus* [9] and strengthen the ability of yellow catfish (*Pelteobagrus fulvidraco*) to resist *Aeromonas hydrophila* infection [10]. Additionally, baicalin yeast culture improved the growth performance, antioxidant capacity, and immune function in giant grouper (*Epinephelus lanceolatus*) [11].

The yellow catfish is an important freshwater aquaculture species in China, valued for its high market demand and nutritional quality. However, in intensive aquaculture systems, farmed fish are generally susceptible to physiological challenges such as oxidative stress and inflammatory responses due to high stocking density and environmental fluctuations. The carnivorous nature and high dietary lipid requirements in this fish suggest a heightened potential for metabolic disorders such as hepatic steatosis. These concerns highlight the need for safe and effective feed additives to enhance growth and health management. Although previous studies have demonstrated that active compounds in Chinese herbal medicine, such as emodin [12], berberine [13], *Ganoderma lucidum* polysaccharides [14], and *Astragalus* polysaccharides [15], could enhance the growth performance and immune function or modulate intestinal microbiota composition in yellow catfish, their application ranges and functional scope remain limited. There is continued interest in identifying novel additives with broader efficacy and better safety profiles. Therefore, this study aims to explore the effects of dietary supplementation of baicalin on growth performance, morphometric indices, serum biochemical parameters, hepatic and intestinal antioxidant capacity, liver health, and intestinal health and microbiota in yellow catfish. The findings will guide the application of baicalin in aquatic feeds.

2. Materials and Methods

All experimental animal care protocols underwent approval by Shanghai Ocean University's Institutional Animal Care and Use Committee (Approval No. SHOU-DW-2024-140). Experimental operations were rigorously conducted in compliance with the standardized ethical regulations promulgated by Shanghai Ocean University's Animal Ethics Oversight Committee.

2.1. Experimental Design and Diets

Five isonitrogenous (46% crude protein) and isolipidic (8.9% crude lipid) diets were formulated using fish meal, soybean meal, and chicken meal as protein sources. According to the previous studies [7,8,10,16], baicalin (purity: 95%, purchased from Macklin Biochemical Technology Co., Ltd., Shanghai, China) was added at concentrations of 0, 100, 200, 400, and 800 mg/kg (BA0, BA100, BA200, BA400, and BA800), respectively. Accordingly, the wheat flour inclusion was reduced to balance the formulation, while the proportions of the other ingredients remained constant. All feed ingredients were ground to pass through a 60-mesh sieve, then thoroughly mixed and extruded to form slow-sinking pellets (2.0 mm diameter) using a single-screw extruder (LX-75, Longxiang Food Machinery Factory, Xingtai, China) with a pelleting temperature of 90 °C. The pellets were air-dried at 60 °C and stored in a cool, dry environment. The diet formulations and nutritional compositions are shown in Table 1.

Table 1. Formulation and proximate composition of experimental diets (air dry basis, %).

Ingredients	BA0	BA100	BA200	BA400	BA800
Fish meal	18.00	18.00	18.00	18.00	18.00
Chicken meal	12.00	12.00	12.00	12.00	12.00
Soybean meal	20.00	20.00	20.00	20.00	20.00
Soybean protein concentrate	6.00	6.00	6.00	6.00	6.00
Wheat flour	22.96	22.95	22.94	22.92	22.88
Corn gluten meal	8.00	8.00	8.00	8.00	8.00
Cottonseed gluten meal	4.00	4.00	4.00	4.00	4.00
Fish oil	1.00	1.00	1.00	1.00	1.00
Soybean oil	2.00	2.00	2.00	2.00	2.00
Soy lecithin	1.30	1.30	1.30	1.30	1.30
Baicalin	0.00	0.01	0.02	0.04	0.08
$Ca(H_2PO_4)$	2.50	2.50	2.50	2.50	2.50
Vitamin and mineral premix [1]	2.00	2.00	2.00	2.00	2.00
Choline chloride	0.20	0.20	0.20	0.20	0.20
Vitamin C phosphate	0.04	0.04	0.04	0.04	0.04
Total	100.00	100.00	100.00	100.00	100.00
Proximate composition [2]					
Crude protein	46.53	46.60	46.71	46.52	46.49
Crude lipid	8.95	8.98	8.97	8.97	8.97
Crude ash	8.63	8.98	8.52	8.90	8.69
Moisture	8.36	8.40	8.61	8.33	8.58

[1] Vitamin and mineral content in premix per kilogram: VA 200,000 IU, VD_3 12,000 IU, VE 1600 mg, VK_3 930 mg, VB_1 280 mg, VB_2 380 mg, VB_6 350 mg, VB_{12} 0.02 mg, and nicotinamide 2500 mg; D-calciumpantothenate 1200 mg, folic acid 2 mg, inositol 100 mg, and biotin 0.2 mg; Fe 540 mg, Zn 830 mg, Mn 500 mg, Cu 200 mg, Co 1 mg, Se 0.35 mg, and I 1 mg. [2] Nutrient levels were all measured values.

2.2. Experimental Fish and Feeding Management

A total of 800 yellow catfish (initial average body weight: 4.15 g) were obtained from Suyu Aquaculture Company (Huzhou, China) and then fed with the control diet for 4 weeks in cement pools at the Binhai Aquaculture Base of Shanghai Ocean University (Shanghai, China). After the acclimation, 225 healthy fish (average weight: 11.19 ± 0.07 g) were selected and randomly stocked into 15 cages (1.0 m × 1.0 m × 1.2 m; mesh size: 0.3 mm; material: Nylon; 15 fish per cage) with 5 treatments and 3 replicates per treatment. Three independent pools with the same water source and management were employed, and each pool contained five cages, which were randomly assigned to the five treatments.

The feeding protocol was conducted with reference to the description by Xu et al. [16] and strictly controlled to ensure consistency across all experimental groups. The daily ration was set at 3% to 6% of the total fish body weight per cage and was adjusted weekly based on collective weighing. Every morning, the feed was weighed to ensure the same amount for each treatment group. The slow-sinking feed was offered by hand twice daily (08:00 and 18:00). To minimize waste and ensure precise intake, the allotted feed for each cage was divided into small batches and introduced manually over 10 min, which has been determined through the pre-trial observation. The uneaten feed settling on the bottom was carefully siphoned out within 30 min and then dried and weighed to calculate the actual feed consumption.

Water quality was detected as follows: temperature 25–30 °C, dissolved oxygen > 5 mg/L, pH 7.5–8.0, ammonia nitrogen < 0.2 mg/L, and nitrite < 0.1 mg/L. One-third of the water was replaced twice weekly with filtered pond water. The feeding trial was conducted at the Binhai Aquaculture Base of Shanghai Ocean University for 56 days.

2.3. Sample Collection

At the end of the feeding trial, the fish were starved for 24 h, and then total weight and fish number per cage were recorded to calculate survival, weight gain (WG), specific growth rate (SGR), and feed conversion ratio (FCR). Before sampling, fish were anesthetized with MS-222 anesthetic. Three fish per cage were randomly selected and stored at −20 °C for whole-body proximate composition analysis. Another three fish were measured for body length and weight to calculate condition factors (CFs). Subsequently, blood (~1 mL) was collected from the caudal vein of the largest three individuals (45–55 g) from each cage using a 1 mL medical syringe. Serum was centrifuged at 3000 rpm for 10 min; then, approximately 1.3 mL of serum was collected per replicate and stored at −80 °C for biochemical assays. After sampling blood, the fish were dissected, and then the viscera and liver weights were recorded to calculate the viscerosomatic index (VSI) and hepatosomatic index (HSI). Liver and intestinal tissues were stored at −80 °C for antioxidant analysis. Liver and intestinal samples from BA0, BA400, and BA800 groups were fixed in Bouin's solution for H&E staining and histological evaluation. Three fish per cage were quickly dissected, and the hind intestinal contents from the BA0 and BA400 groups were collected aseptically and placed into enzyme-free freezing tubes for preservation at −80 °C to analyze the microbiota community.

2.4. Measurement Indicators and Methods

2.4.1. Growth Performance and Body Morphometric Indices

$$\text{SR}\ (\%) = (\text{final number of fish}/\text{initial number of fish}) \times 100$$

$$\text{WG}\ (\%) = [(\text{final body weight} - \text{initial body weight})/\text{initial body weight}] \times 100$$

$$\text{SGR}\ (\%) = 100 \times [\ln(\text{final body weight}) - \ln(\text{initial body weight})]/\text{experiment days}$$

$$\text{FCR} = \text{feed intake (g)}/\text{weight gain (g)}$$

$$\text{HSI}\ (\%) = [\text{final liver weight (g)}/\text{final body weight (g)}] \times 100$$

$$\text{VSI}\ (\%) = [\text{final visceral weight (g)}/\text{final body weight (g)}] \times 100$$

$$\text{CF} = \text{final body weight (g)}/\text{body length (cm)}^3 \times 100$$

2.4.2. Proximate Composition of Diets and Whole Body

Referring to the AOAC method, the moisture content in feed and whole fish was determined by oven drying at 105 °C, and crude protein was analyzed using the Kjeldahl nitrogen analyzer (2300-Auto-Analyzer, Foss Tecator, Höganäs, Sweden), while ash content was quantified by combustion at 550 °C in a muffle furnace (SXL-1008, Shanghai Jinhong Experimental Equipment Co., Shanghai, China). Crude fat was measured by the chloroform–methanol assay [17].

2.4.3. Serum, Hepatic, and Intestinal Biochemical Parameters

The liver and intestinal tissues were homogenized at 10,000 rpm for 10 s in ice-cold physiological saline (1:9 w/v) and then centrifuged at 3000 rpm for 10 min (4 °C). The supernatants were collected and stored at −80 °C for subsequent analyses.

The measurements of total cholesterol (TCHO; COD-PAP assay), triglycerides (TGs; GPO-PAP assay), total protein (TP; Coomassie brilliant blue assay), lysozyme (LZM; turbidimetric assay), alkaline phosphatase and acid phosphatase (ALP and ACP; microplate enzymatic assay), alanine aminotransferase and aspartate aminotransferase (ALT and AST; microplate colorimetric assay), total antioxidant capacity (T-AOC; FRAP assay), malondialdehyde (MDA; TBA assay), catalase (CAT; ammonium molybdate colorimetry assay), and superoxide dismutase (SOD; hydroxylamine oxidation assay) were performed using commercial kits (Nanjing Jiancheng Bioengineering Institute, Nanjing, China).

2.4.4. Histological Examination

The liver and intestinal tissues from the BA0, BA400, and BA800 groups underwent ethanol dehydration, xylene clearing, and paraffin embedding. Sectioned samples were stained with hematoxylin–eosin (H&E) and mounted for microscopic observation. Histomorphological features were captured using optical microscopy (Olympus DP71), and the vacuolization area quantification and intestinal morphology measurements were performed through ImageJ 1.53 software.

2.4.5. Intestinal Microbiota Analysis Based on 16S rRNA

Intestinal content DNA from the BA0 and BA400 groups was extracted using QIAamp DNA Stool Mini Kit (Qiagen, Venlo, The Netherlands). After purity verification (NanoDrop 2000, Thermo Scientific, Wilmington, DE, USA), the V3–V4 hypervariable regions of bacterial 16S rRNA genes were amplified and sequenced on the Illumina NovaSeq 6000 platform (2×250 bp paired-end reads). Raw sequences were processed through the QIIME2 pipeline with DADA2 denoising, chimera removal, and 100% similarity clustering. Operational taxonomic units (OTUs) were taxonomically classified against the SILVA 138 database. Alpha diversity indices and KEGG functional predictions were generated using PICRUSt2. Sequencing services were provided by Shanghai Shunshi Biotechnology Co., Ltd. (Shanghai, China).

2.5. Statistical Analysis

All data were expressed as mean ± standard deviation (SD). One-way ANOVA was performed using SPSS 26.0. Tukey's HSD test was used for multiple comparisons. A p-value below 0.05 was considered statistically significant. Graphical visualizations were created with GraphPad Prism 9.0 (GraphPad Software) and Microsoft Excel 2021.

3. Results

3.1. Growth Performance and Morphometric Indices

Compared with the control group (BA0), the final weight, survival, WG, and SGR of the BA200 and BA400 groups were significantly higher ($p < 0.05$), along with a significantly lower FCR ($p < 0.05$). No significant differences were observed in growth performance among the BA100, BA800, and control (BA0) groups ($p > 0.05$) except for survival. CF, VSI, and HSI showed no significant differences among all the groups ($p > 0.05$) (Table 2).

Table 2. Dietary effects of baicalin on growth performance and body parameters of *Pelteobagrus fulvidraco*.

Items	BA0	BA100	BA200	BA400	BA800
Initial weight (g)	11.23 ± 0.05	11.20 ± 0.04	11.20 ± 0.07	11.27 ± 0.08	11.09 ± 0.04
Final weight (g)	31.82 ± 0.25 ^{a}	32.57 ± 0.80 ^{a}	36.30 ± 1.14 ^{bc}	39.10 ± 1.08 ^{c}	33.23 ± 1.27 ^{ab}
Survival (%)	95.56 ± 3.85 ^{a}	100 ± 0.00 ^{b}	100 ± 0.00 ^{b}	100 ± 0.00 ^{b}	100 ± 0.00 ^{b}
WG (%)	183.28 ± 3.44 ^{a}	190.77 ± 7.16 ^{a}	228.79 ± 11.14 ^{bc}	247.04 ± 9.62 ^{c}	199.43 ± 12.74 ^{ab}
SGR (%)	1.75 ± 0.05 ^{a}	1.88 ± 0.06 ^{a}	2.12 ± 0.06 ^{bc}	2.22 ± 0.05 ^{c}	1.96 ± 0.08 ^{ab}
FCR	1.80 ± 0.02 ^{a}	1.71 ± 0.07 ^{a}	1.42 ± 0.06 ^{b}	1.32 ± 0.06 ^{b}	1.66 ± 0.09 ^{a}
CF (g/cm^3)	1.75 ± 0.28	1.84 ± 0.16	1.74 ± 0.15	1.82 ± 0.10	1.81 ± 0.29
VSI (%)	8.83 ± 0.82	7.87 ± 2.56	8.46 ± 1.18	7.93 ± 0.60	8.86 ± 1.44
HSI (%)	2.56 ± 0.33	2.52 ± 0.10	2.31 ± 0.32	2.36 ± 0.16	2.34 ± 0.29

Data with different letters in the same row indicate a significant difference ($p < 0.05$). The same applies to the tables below.

3.2. Serum Biochemical Indices

In Table 3, serum TG levels in the BA100 and BA400 groups, as well as AST activity across all baicalin-supplemented groups, were significantly lower than those in the control group (BA0) ($p < 0.05$). The BA400 group displayed significantly elevated ALP activity, while both the BA200 and BA400 groups showed higher LZM activity than the control ($p < 0.05$). No significant differences were detected in TCHO, TP content, or ALT activity among all the groups ($p > 0.05$).

Table 3. Dietary effects of baicalin on serum biochemical indices of *Pelteobagrus fulvidraco*.

Items	BA0	BA100	BA200	BA400	BA800
TCHO (mmol/L)	8.11 ± 0.80	7.04 ± 0.04	7.68 ± 0.33	6.99 ± 0.52	7.95 ± 0.55
TG (mmol/L)	10.68 ± 1.11 ^{ab}	9.11 ± 0.41 ^{c}	9.41 ± 0.79 ^{bc}	9.19 ± 0.79 ^{c}	10.87 ± 0.84 ^{a}
ALP (U/L)	67.32 ± 3.46 ^{a}	66.51 ± 6.25 ^{a}	73.94 ± 6.64 ^{a}	90.43 ± 8.73 ^{b}	74.06 ± 5.66 ^{a}
ACP (U/L)	286.03 ± 17.10 ^{a}	296.00 ± 18.12 ^{ab}	299.24 ± 17.66 ^{ab}	314.12 ± 15.11 ^{b}	304.47 ± 17.51 ^{ab}
ALT (U/L)	4.39 ± 0.32	4.67 ± 0.19	4.41 ± 0.16	4.55 ± 0.27	4.46 ± 0.14
AST (U/L)	140.38 ± 11.42 ^{a}	101.43 ± 14.04 ^{bc}	81.32 ± 1.67 ^{c}	87.63 ± 1.67 ^{c}	110.00 ± 12.38 ^{b}
TP (gprot/L)	31.61 ± 1.10	29.82 ± 2.02	30.58 ± 1.58	30.32 ± 2.52	29.42 ± 1.92
LZM (ug/mL)	1.18 ± 0.19 ^{a}	1.39 ± 0.17 ^{a}	1.79 ± 0.19 ^{b}	2.36 ± 0.23 ^{c}	1.51 ± 0.26 ^{ab}

Data with different letters in the same row indicate a significant difference ($p < 0.05$).

3.3. Antioxidant Capacity

As shown in Table 4, the T-AOC, CAT, and SOD activities in liver and intestine increased firstly and then decreased with the increasing dietary baicalin levels, peaking in the BA400 group with significantly higher values than the control ($p < 0.05$). The BA400 group also presented significantly lower MDA content than the control ($p < 0.05$). Intestinal LZM activity showed no differences among groups ($p > 0.05$).

Table 4. Dietary effects of baicalin on antioxidant indices in the liver and intestine of *Pelteobagrus fulvidraco*.

Items	BA0	BA100	BA200	BA400	BA800
Liver					
T-AOC (mmol/mg prot)	0.16 ± 0.05 [a]	0.23 ± 0.04 [bc]	0.26 ± 0.04 [cd]	0.31 ± 0.05 [d]	0.19 ± 0.04 [ab]
MDA (nmol/mgprot)	4.44 ± 0.58 [a]	2.96 ± 0.76 [b]	2.50 ± 0.36 [b]	2.35 ± 0.58 [b]	2.91 ± 0.41 [b]
CAT (U/mgprot)	8.70 ± 1.91 [a]	11.93 ± 2.26 [ab]	13.00 ± 1.90 [bc]	15.71 ± 2.31 [c]	11.56 ± 2.26 [ab]
SOD (U/mgprot)	71.81 ± 10.11 [a]	94.69 ± 10.01 [b]	109.58 ± 4.38 [c]	140.86 ± 5.33 [d]	109.74 ± 6.08 [c]
Intestine					
T-AOC (μmol/mg prot)	0.017 ± 0.004 [a]	0.022 ± 0.003 [ab]	0.026 ± 0.004 [b]	0.025 ± 0.004 [b]	0.019 ± 0.003 [ab]
MDA (nmol/mgprot)	1.50 ± 0.26 [a]	0.76 ± 0.20 [b]	0.72 ± 0.14 [b]	0.79 ± 0.19 [b]	1.17 ± 0.28 [ab]
CAT (U/mgprot)	18.08 ± 1.58 [a]	25.55 ± 5.08 [ab]	23.71 ± 4.98 [ab]	31.21 ± 3.76 [b]	20.84 ± 3.36 [a]
LZM (ug/mgprot)	0.06 ± 0.01	0.064 ± 0.01	0.060 ± 0.01	0.079 ± 0.009	0.070 ± 0.01
SOD (U/mgprot)	6.99 ± 2.38 [a]	8.12 ± 2.39 [a]	14.43 ± 2.30 [b]	16.56 ± 1.25 [b]	9.84 ± 2.93 [a]

Data with different letters in the same row indicate a significant difference ($p < 0.05$).

3.4. Whole-Body Proximate Composition

There were no significant differences in whole-body composition among all groups ($p > 0.05$), including moisture, crude ash, crude lipid, and crude protein contents (Table 5).

Table 5. Dietary effects of baicalin on the body proximate composition of *Pelteobagrus fulvidraco* (wet weight, %).

Items	BA0	BA100	BA200	BA400	BA800
Moisture	72.38 ± 0.40	72.12 ± 1.08	72.25 ± 0.70	72.00 ± 1.09	71.65 ± 0.78
Crude ash	4.06 ± 0.53	4.00 ± 0.23	4.08 ± 0.21	3.95 ± 0.05	4.07 ± 0.20
Crude lipid	6.48 ± 0.46	6.61 ± 0.46	6.51 ± 0.36	6.46 ± 0.59	6.37 ± 0.45
Crude protein	15.33 ± 0.64	15.64 ± 0.68	15.33 ± 0.73	15.60 ± 0.67	15.89 ± 0.26

3.5. Liver and Intestinal Histomorphology

In Figure 1, the BA0 group exhibited marked hepatic vacuolization (indicated by black arrows), whereas hepatocytes in the BA400 group were densely arranged with normal morphology. The moderate vacuolization was also observed in the BA800 group. Quantitative analysis revealed significantly reduced hepatic vacuolization area in the BA400 group compared to the BA0 group ($p < 0.05$).

The intestinal villi in the BA400 and BA800 groups were orderly arranged with compact texture compared to the BA0 group (Figure 2). In Table 6, the BA400 group presented significantly greater villus height, villus width, and muscularis thickness than the control ($p < 0.05$), while BA800 displayed no significant differences ($p > 0.05$).

Table 6. Dietary effects of baicalin on intestinal tissue morphology of *Pelteobagrus fulvidraco*.

Items (μm)	BA0	BA400	BA800
Villus height	570.4 ± 20.13 [a]	670.9 ± 51.51 [b]	561.2 ± 31.65 [a]
Villus width	105.6 ± 5.44 [a]	138.4 ± 19.65 [b]	120.8 ± 14.78 [ab]
Muscularis thickness	167.5 ± 26.05 [a]	222.6 ± 15.29 [b]	197.1 ± 23.30 [ab]

Data with different letters in the same row indicate a significant difference ($p < 0.05$).

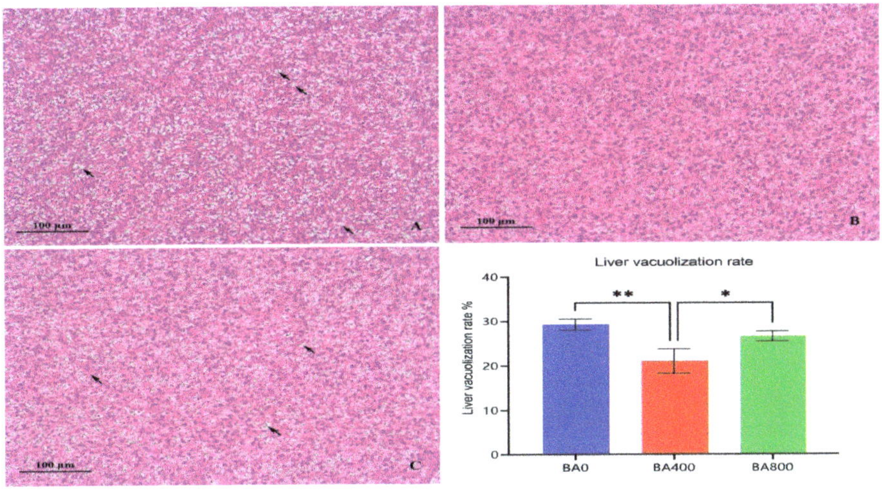

Figure 1. Liver tissue sections and liver vacuolization rate of *Pelteobagrus fulvidraco*. (**A–C**) represent the three groups of BA0, BA400, and BA800. * Significant difference between two groups at $p < 0.05$; ** significant difference between two groups at $p < 0.01$. Black arrows indicate hepatic vacuolization.

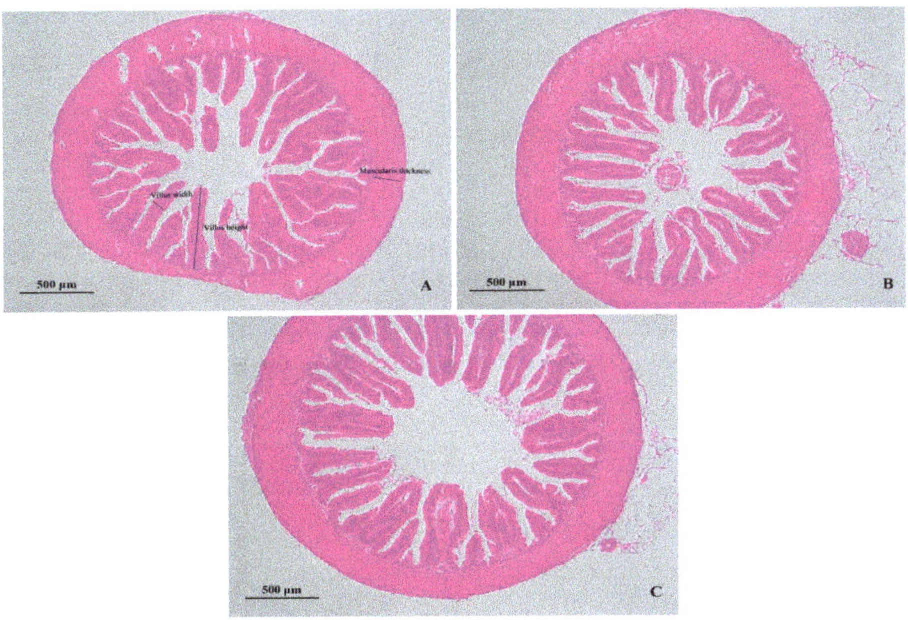

Figure 2. Intestine tissue sections of *Pelteobagrus fulvidraco*; (**A–C**) represent the three groups of BA0, BA400, and BA800.

3.6. Intestinal Microbiota Analysis

The coverage values for all groups approached 1, indicating that the microbial community was sufficiently sampled and the obtained data accurately represented the underlying microbial population. As shown in Table 7, the Shannon and Simpson indices of intestinal microbiota in the BA400 group were significantly higher than those in the BA0 group ($p < 0.05$), while no significant differences were observed in Chao1 and ACE (Abundance-based Coverage Estimator) indices between the two groups ($p > 0.05$). A total of 1155 OTUs were identified in the intestinal contents from both groups, with 124 OTUs shared by the two groups, and the BA0 and BA400 groups exhibited 518 and 513 unique OTUs, respectively (Figure 3A).

Table 7. Dietary effects of baicalin on the alpha diversity index of intestinal microbiota of Pelteobagrus fulvidraco.

Items	BA0	BA400
Shannon	3.01 ± 0.12^{a}	4.69 ± 0.007^{b}
Simpson	0.64 ± 0.03^{a}	0.89 ± 0.006^{b}
Chao1	265.53 ± 12.58	299.22 ± 26.85
ACE	266.25 ± 12.24	299.70 ± 26.89
Coverage	0.999 ± 0.00	0.999 ± 0.00

Data with different letters in the same row indicate a significant difference ($p < 0.05$).

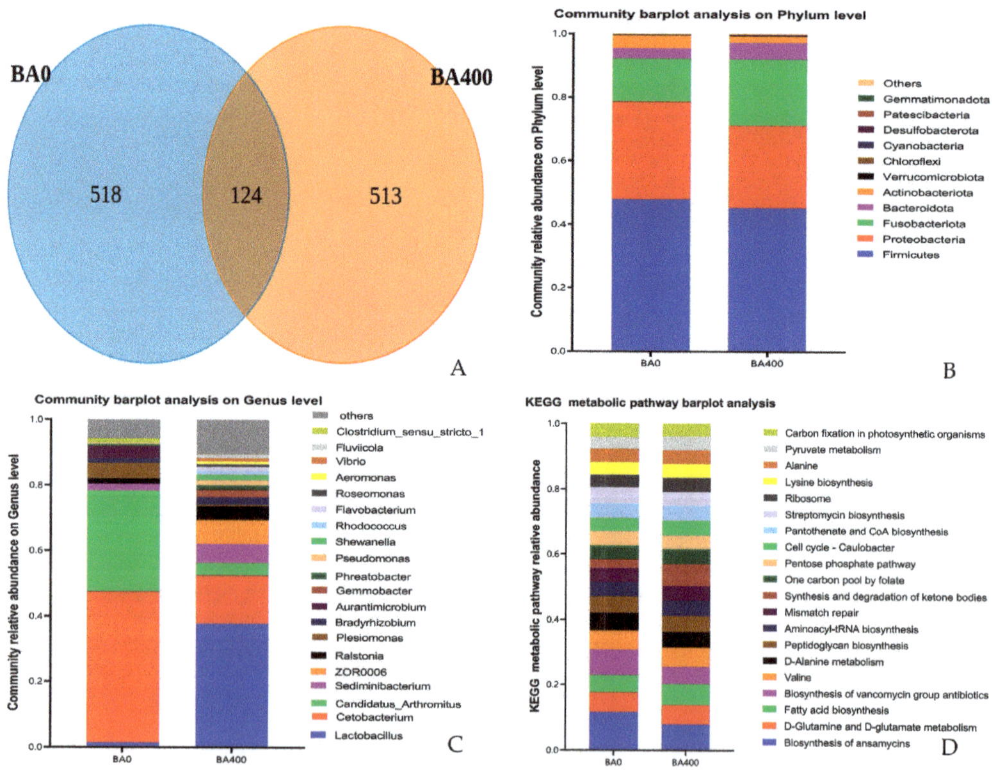

Figure 3. Dietary effects of baicalin on intestinal microbiota of *Pelteobagrus fulvidraco*. (**A**) Venn diagram of OTUs. (**B**) Community barplot analysis on the phylum level. (**C**) Community barplot analysis on the genus level. (**D**) KEGG metabolic pathway barplot analysis.

At the phylum level, the dominant phyla in both BA0 and BA400 groups were *Firmicutes*, *Proteobacteria*, and *Fusobacteriota*. Compared with the BA0 group, the BA400 group showed significantly decreased relative abundances of *Proteobacteria* and *Actinobacteriota* ($p < 0.05$), along with a significant increase in *Fusobacteriota* abundance ($p < 0.05$) (Figure 3B).

At the genus level, the dominant genera in the BA0 group were *Cetobacterium*, *Candidatus_Arthromitus*, and *Plesiomonas*, whereas the BA400 group was dominated by *Lactobacillus*, *Cetobacterium*, and *ZOR0006*. Significant increases in the relative abundances of *Lactobacillus* and ZOR0006 were observed in the BA400 group compared to the BA0 group ($p < 0.05$), while the abundances of *Cetobacterium*, *Candidatus_Arthromitus*, and *Plesiomonas* were significantly reduced ($p < 0.05$) (Figure 3C).

The BA400 group exhibited a significantly enhanced metabolic potential of fatty acid synthesis and ketone body synthesis compared to the BA0 group ($p < 0.05$). Conversely, the synthesis pathways of ansamycins and vancomycin-class antibiotics were significantly downregulated in the BA400 group ($p < 0.05$) (Figure 3D).

4. Discussion

4.1. Growth Performance

Jia et al. [7] reported that 400 mg/kg dietary baicalin supplementation significantly decreased the FCR in Nile tilapia (*Oreochromis niloticus*). As a structurally analogous flavonoid to baicalin, baicalein (200 mg/kg) was shown to enhance the growth performance of grass carp (*Ctenopharyngodon idella*) [16]. In hybrid grouper (*Epinephelus fuscoguttatus*♀× *E. lanceolatus*♂), dietary supplementation with 1–3% *Scutellaria baicalensis* extract (containing baicalin) significantly increased the growth, which might be attributed to the upregulation of growth hormone receptor (GHR), insulin-like growth factor I (IGF-I), and IGF-II gene expression, along with the enhanced intestinal digestive enzyme activity and antioxidant capacity [18]. Similarly, 10 g/kg *Scutellaria baicalensis* (the botanical source of baicalin) supplementation decreased the FCR of channel catfish [19]. Consistent with these findings, our trial revealed that 200–400 mg/kg baicalin addition significantly increased WG and reduced FCR in yellow catfish.

Notably, the 800 mg/kg baicalin group showed significantly reduced growth performance compared to the BA400 group, suggesting potential toxicological effects at high doses. This observation aligns with the reports that excessive flavonoid intake may induce pro-oxidant toxicity through free radical-mediated DNA damage or inhibition of DNA-associated enzymes (e.g., topoisomerases) [20], while causing DNA and chromosomal mutations as a mutagen [21]. Notably, the significantly increased feed conversion ratio (FCR) at the 800 mg/kg supplementation level indicates a metabolic impairment in nutrient utilization rather than a taste-mediated avoidance behavior. Similar dose-dependent growth suppression has also been encountered in channel catfish fed with a 15 g/kg *Scutellaria baicalensis* addition [19] and grass carp fed with a 600 mg/kg baicalein addition [17].

4.2. Serum, Liver, and Intestinal Biochemical Indices

Alkaline phosphatase (ALP) regulates immune function by promoting the dephosphorylation of bacterial or endogenous pro-inflammatory factors [22]. Acid phosphatase (ACP), primarily located in lysosomes, modulates immunity by dephosphorylating pathogenic substances engulfed by lysosomes [23]. Lysozyme (LZM) exerts bactericidal effects by dissolving peptidoglycan in bacterial cell walls [24]. Catalase (CAT) and superoxide dismutase (SOD) efficiently scavenge free radicals in organisms [23]. Malondialdehyde (MDA), a terminal product of lipid peroxidation, reflects the severity of oxidative stress on biomembrane systems and serves as a critical indicator of oxidative damage [25]. Triglyceride (TG) is a key parameter for assessing blood lipid levels and plays vital roles in lipid metabolism and energy homeostasis in fish.

Baicalin exerts anti-inflammatory and immuno-modulatory effects by regulating the TLRs/MyD88/NF-κB and TLRs/MyD88/MAPK signaling pathways [26]. Bao et al. [2] reported that baicalin combats viral infections through direct virucidal activity, inhibition of viral replication, modulation of host cell functional protein expression, and anti-inflammatory mechanisms, while also showing antibacterial and antifungal properties to enhance disease resistance ability. Previous studies have demonstrated that baicalin protected Nile tilapia infected with *Streptococcus agalactiae* by attenuating bacterial virulence and enhancing non-specific immunity [8] and improved the survival of yellow catfish challenged with *Aeromonas hydrophila* [10]. Furthermore, Li et al. [9] found that baicalin upregulated the expression of crustin, anti-lipopolysaccharide factors (ALFs), and LZM genes in *Macrobrachium rosenbergii* infected with *Vibrio parahaemolyticus*. In this experiment, baicalin significantly increased serum ALP, ACP, and LZM activities, indicating its capacity to enhance non-specific immune responses in yellow catfish.

Baicalin exhibits strong antioxidant activity due to the two phenolic hydroxyl groups, which can scavenge free radicals and reactive oxygen species [27]. Additionally, baicalin can reduce oxidative stress by modulating the Nrf2-Keap1 and p38 MAPK signaling pathways in the liver [2]. Jia et al. [7] demonstrated that baicalin alleviated H_2O_2-induced oxidative stress in Nile tilapia by elevating hepatic SOD, total antioxidant capacity (T-AOC), and glutathione (GSH) levels. Yan et al. [10] reported that intraperitoneal injection of baicalin enhanced SOD, GSH, and CAT activities in the liver of yellow catfish infected with *A. hydrophila*. In the present study, baicalin increased T-AOC, CAT, and SOD levels while reducing MDA content in both liver and intestinal tissues, confirming its antioxidant-enhancing properties.

Dietary supplementation with 400 mg/kg baicalin mitigated obesity and hyperlipidemia in mice fed with high-fat diets [28]. Elshopakey et al. [29] observed that baicalin significantly reduced serum triglyceride levels in Nile tilapia, and *Scutellaria baicalensis* extract also decreased serum triglycerides in Far Eastern catfish (*Silurus asotus*) [30]. Consistent with these findings, the present results demonstrated that baicalin effectively lowered serum triglyceride levels in yellow catfish. The reduction is considered beneficial for fish health, as it helps maintain metabolic homeostasis and reduces the risk of lipid-related disorders like hepatic steatosis, as we mentioned below.

4.3. Liver and Intestinal Histology

Hepatic vacuolization, typically resulting from excessive lipid accumulation, may impair the normal metabolic functions [31]. Aspartate aminotransferase (AST), a key transaminase in amino acid metabolism, predominantly localizes in the liver and exhibits elevated serum concentrations during hepatic dysfunction [32]. Baicalin could reduce hepatic lipid deposition in mice by activating AMP-activated protein kinase (AMPK) and acetyl-CoA carboxylase (ACC), while downregulating lipogenesis-related genes, including fatty acid synthase and its upstream regulator SREBP-1c [33]. It has been reported that baicalin decreased H_2O_2-induced hepatic vacuolization in Nile tilapia [7], and dietary supplementation with *Scutellaria baicalensis* reduced serum AST activity and hepatic vacuolization in channel catfish [19]. In this study, 400 mg/kg baicalin supplementation significantly reduced hepatic vacuolization and serum AST activity in yellow catfish, demonstrating its hepato-protective effects.

Intestinal morphological parameters, including villus width, villus height, and muscularis thickness, directly reflect intestinal health. Villus width and height are closely associated with nutrient absorption capacity [34]. Baicalin has been shown to protect colonic tissue by inhibiting intestinal epithelial cell apoptosis and enhancing tight junction protein expression through suppression of the PI3K/AKT signaling pathway [35]. Zhang et al. [36] reported that dietary supplementation with a *Scutellaria*-containing herbal mixture significantly increased intestinal villus width and height in large yellow croaker (*Larimichthys crocea*). The current experiment revealed that 400 mg/kg baicalin supplementation significantly improved intestinal villus width, villus height, and muscularis thickness in yellow catfish, confirming its efficacy in enhancing intestinal structural integrity.

4.4. Intestinal Microbiota

The intestinal microbiota is a vital component of the gut, playing significant roles in nutrient absorption, immune system development, and defense against pathogenic invasions [37]. The biodiversity and community composition of the intestinal microbiota can reflect intestinal health [38]. In this experiment, baicalin significantly enhanced the alpha diversity (Shannon and Simpson indices) of the intestinal microbiota in yellow catfish. At the phylum level, the dominant phyla in both groups were *Firmicutes*, *Proteobacteria*, and *Fusobacteria*, consistent with the previous findings in yellow catfish [39]. Compared to the control group, the baicalin-supplemented group exhibited significantly reduced relative abundances of *Proteobacteria* and *Actinobacteria* and significantly increased the abundance of *Fusobacteria*. *Proteobacteria* contain some pathogenic and opportunistic pathogenic genera that may impair intestinal health in fish [40]. In contrast, *Fusobacteria* can synthesize various vitamins and short-chain fatty acids (SCFAs) [41,42], potentially promoting intestinal health for yellow catfish.

At the genus level, the dominant genera in the control group were *Cetobacterium*, *Candidatus_Arthromitus*, and *Plesiomonas*, whereas the baicalin-supplemented group (400 mg/kg) showed dominance of *Lactobacillus*, *Cetobacterium*, and *ZOR0006*. These results aligned with the previous reports by Chen et al. [39], who identified *Cetobacterium* and *Candidatus_Arthromitus* as dominant genera in yellow catfish, and by Zhu et al. [43], who reported *Cetobacterium* and *Plesiomonas* as key genera. Compared to the control, baicalin supplementation significantly increased the relative abundances of *Lactobacillus* and *ZOR0006*, while reducing the relative abundances of *Cetobacterium*, *Candidatus_Arthromitus*, and *Plesiomonas*. *Lactobacillus*, known as a probiotic in fish intestines, is able to regulate the intestinal environment via lactic acid secretion [44]. *ZOR0006*, an intestinal-specific bacterium, participates in carbohydrate decomposition and lactate synthesis [45]. Yan et al. [10] demonstrated that baicalin elevated Chao1 and ACE indices of intestinal microbiota in yellow catfish infected with *Aeromonashydrophila*. Xia et al. [46] found that *Scutellaria baicalensis* extract reduced the harmful *Deltaproteobacteria* level in the intestine of rabbit fish (*Siganus fuscescens*). Similarly, Du et al. [47] reported that baicalin improved intestinal microbial composition in koi carp. Consistent with these previous findings, the present results indicated that baicalin enhanced intestinal microbial richness and positively modulated microbiota composition in yellow catfish.

Analysis of KEGG metabolic pathways derived from intestinal microbiota composition reflects microbial metabolic activity [48]. The fatty acid synthesis in the gut is primarily driven by anaerobic bacterial fermentation of carbohydrates to produce SCFAs [49], which can inhibit harmful bacteria and regulate host immunity [50]. Baicalin has been shown to enhance SCFA synthesis in mice by increasing the abundances of *Firmicutes*, *Bacteroidetes*, and *Fusobacteria* [51]. Xia et al. [46] also observed the elevated SCFA synthesis in the intestines of rabbit fish fed with a *Scutellaria baicalensis* extract-supplemented diet. In this study, the increased fatty acid metabolism in the baicalin group may correlate with the higher abundances of *Fusobacteria* and *Bacteroidetes*. Additionally, *Actinobacteria* can reduce microbial diversity and inhibit probiotic growth through the production of antibiotics, such as ansamycins and vancomycins [52–54]. The decreased synthesis of these antibiotics in the present baicalin group likely corresponded to the reduced *Actinobacteria* abundance.

5. Conclusions

Dietary supplementation of baicalin significantly improved weight gain, feed utilization, antioxidant capacity, and liver and intestinal tissue morphology in yellow catfish, while positively modulating intestinal microbiota composition. Under the experimental conditions, the optimal dietary supplementation level of baicalin is 400 mg/kg for yellow catfish. These benefits suggest that baicalin could serve as a functional feed additive to promote fish health, growth, and potentially improve product quality in farmed yellow catfish.

Future research should focus on validating the positive effects in practical farming environments and exploring the molecular mechanisms underlying baicalin's regulatory roles in gut–liver crosstalk, immunity, and nutrient metabolism.

Author Contributions: Conceptualization, X.L. (Xiangjun Leng) and H.L.; methodology, H.L., X.L. (Xinru Li) and Y.C.; validation, X.L. (Xinru Li) and Y.F.; formal analysis, H.L.; investigation, H.L., X.L. (Xinru Li), Y.F. and Y.X.; data curation, H.L.; writing—original draft, H.L.; writing—review and editing, X.L. (Xiangjun Leng); visualization, H.L.; supervision, X.L. (Xiangjun Leng) and X.L. (Xiaoqin Li); project administration, X.L. (Xiangjun Leng) and X.L. (Xiaoqin Li); funding acquisition, X.L. (Xiangjun Leng). All authors have read and agreed to the published version of the manuscript.

Funding: This research was funded by the National Key R&D Program of China (2024YFD2402005).

Institutional Review Board Statement: All experimental animal care protocols underwent approval by Shanghai Ocean University's Institutional Animal Care and Use Committee (Approval No. SHOU-DW-2024-140). Experimental operations were rigorously conducted in compliance with the standardized ethical regulations promulgated by Shanghai Ocean University's Animal Ethics Oversight Committee.

Informed Consent Statement: Not applicable.

Data Availability Statement: Data are contained within the article.

Conflicts of Interest: The authors declare no conflicts of interest.

References

1. Wang, Z.-L.; Wang, S.; Kuang, Y.; Hu, Z.-M.; Qiao, X.; Ye, M. A comprehensive review on phytochemistry, pharmacology, and flavonoid biosynthesis of *Scutellaria baicalensis*. *Pharm. Biol.* **2018**, *56*, 465–484. [CrossRef]
2. Bao, M.; Ma, Y.; Liang, M.; Sun, X.; Ju, X.; Yong, Y.; Liu, X. Research progress on pharmacological effects and new dosage forms of baicalin. *Vet. Med. Sci.* **2022**, *8*, 2773–2784. [CrossRef]
3. Hu, F.; Bi, Y.; Zheng, X.; Lu, M.; Diao, Q.; Tu, Y. Effect of baicalin supplementation on the growth, health, antioxidant and anti-inflammatory capacity, and immune function of preweaned calves. *Anim. Feed. Sci. Technol.* **2023**, *298*, 115598. [CrossRef]
4. Liu, X.; Ji, Y.; Miao, Z.; Lv, H.; Lv, Z.; Guo, Y.; Nie, W. Effects of baicalin and chlorogenic acid on growth performance, slaughter performance, antioxidant capacity, immune function and intestinal health of broilers. *Poult. Sci.* **2024**, *103*, 104251. [CrossRef] [PubMed]
5. Chen, Y.; Yang, Y.; Wang, F.; Yang, X.; Yao, F.; Ming, K.; Yuan, W.; Zeng, L.; Liu, J. Antiviral effect of baicalin phospholipid complex against duck hepatitis A virus type 1. *Poult. Sci.* **2018**, *97*, 2722–2732. [CrossRef]
6. Zha, A.; Yuan, D.; Cui, Z.; Qi, M.; Liao, S.; Liao, P.; Tan, B. The evaluation of the antioxidant and intestinal protective effects of baicalin-copper in deoxynivalenol-challenged piglets. *Oxidative Med. Cell. Longev.* **2020**, *2020*, 5363546. [CrossRef] [PubMed]
7. Jia, R.; Du, J.; Cao, L.; Feng, W.; Xu, P.; Yin, G. Effects of dietary baicalin supplementation on growth performance, antioxidative status and protection against oxidative stress-induced liver injury in GIFT tilapia (*Oreochromis niloticus*). *Comp. Biochem. Physiol. Part C Toxicol. Pharmacol.* **2021**, *240*, 108914. [CrossRef] [PubMed]
8. Wang, F.; Xian, X.-R.; Guo, W.-L.; Zhong, Z.-H.; Wang, S.-F.; Cai, Y.; Sun, Y.; Chen, X.-f.; Wang, Y.-q.; Zhou, Y.-C. Baicalin attenuates *Streptococcus agalactiae* virulence and protects tilapia (*Oreochromis niloticus*) from group B streptococcal infection. *Aquaculture* **2020**, *516*, 734645. [CrossRef]

9. Li, X.L.; Zhang, Y.F.; Shen, P.j.; Xu, Y.; Gao, Q.; Cheng, H.H.; Wang, W.D.; Gu, Z.m.; Chen, X.F. Baicalin protects giant freshwater prawn, *Macrobrachium rosenbergii* against *Vibrio parahaemolyticus* via modulation of the Toll signaling pathway. *J. World Aquac. Soc.* **2021**, *52*, 987–1000. [CrossRef]
10. Yan, P.; Liu, J.; Huang, Y.; Yi, T.; Zhang, H.; Dai, G.; Wang, X.; Gao, Z.; He, B.; Guo, W. Baicalin enhances antioxidant, inflammatory defense, and microbial diversity of yellow catfish (*Pelteobagrus fulvidraco*) infected with *Aeromonas hydrophila*. *Front. Microbiol.* **2024**, *15*, 1465346. [CrossRef]
11. Chen, G.-L. *Effects of Baicalin Yeast Culture on Growth, Muscle Composition, Antioxidant Capacity, and Immune Performance in Giant Grouper (*Epinephelus lanceolatus*)*; Southwest University: Chongqing, China, 2020.
12. Liu, F.; Shi, H.-Z.; Guo, Q.-S.; Yu, Y.-B.; Wang, A.-M.; Lv, F.; Shen, W.-B. Effects of astaxanthin and emodin on the growth, stress resistance and disease resistance of yellow catfish (*Pelteobagrus fulvidraco*). *Fish Shellfish. Immunol.* **2016**, *51*, 125–135. [CrossRef] [PubMed]
13. Wang, L.; Gao, C.; Yang, L.; Wang, C.; Wang, B.; Wang, H.; Shu, Y.; Yan, Y. The growth-promoting and lipid-lowering effects of berberine are associated with the regulation of intestinal bacteria and bile acid profiles in yellow catfish (*Pelteobagrus fulvidraco*). *Aquac. Rep.* **2023**, *33*, 101848. [CrossRef]
14. Yao, J.Y.; Xu, Y.; Sheng, P.C.; Yin, W.L.; Lin, L.Y.; Yuan, X.M.; Pan, X.Y.; Hao, G.J.; Shen, J.Y. Effects of three immunoenhancers on non-specific immune function in yellow catfish (*Pelteobagrus fulvidraco*). *J. Anhui Agric. Sci.* **2014**, *42*, 2921–2923.
15. Wu, X.; Bai, D.Q.; Yang, G.; Ning, B.; Zhang, X.T.; Sun, L.Z. Effects of *Ganoderma lucidum* polysaccharides on immune cell activity in yellow catfish (*Pelteobagrus fulvidraco*). *Acta Agric. Boreali-Sin.* **2011**, *26*, 195–198.
16. Xu, Z.; Yang, H.; Liang, G.Y.; Gao, B.W.; Li, X.Q.; Leng, X.J. Effects of baicalein on growth performance, serum antioxidant indices, and muscle quality of grass carp (*Ctenopharyngodon idellus*). *J. Fish. China* **2019**, *43*, 2383–2393.
17. Folch, J.; Lees, M.; Stanley, G.S. A simple method for the isolation and purification of total lipides from animal tissues. *J. Biol. Chem.* **1957**, *226*, 497–509. [CrossRef]
18. Xia, Y.-T.; Wu, Q.-Y.; Cheng, E.H.-C.; Dong, T.T.-X.; Qin, Q.-W.; Wang, W.-X.; Tsim, K.W.-K. The inclusion of extract from aerial part of *Scutellaria baicalensis* in feeding of pearl gentian grouper (*Epinephelus fuscoguttatus*♀× *Epinephelus lanceo-latus*♂) promotes growth and immunity. *Fish Shellfish. Immunol.* **2022**, *127*, 521–529. [CrossRef] [PubMed]
19. Liu, H.N.; Leng, X.J.; You, Y.H.; Liu, J.W.; Chen, Y.F.; Li, X.Q. Effects of *Scutellaria baicalensis* on growth performance, serum and liver biochemical indices, and liver histology of channel catfish (*Ictalurus punctatus*). *J. Dalian Ocean. Univ.* **2024**, *39*, 739–745.
20. Dičkancaitė, E.; Nemeikaitė, A.; Kalvelytė, A.; Čėnas, N. Prooxidant character of flavonoid cytotoxicity: Structure-activity relationships. *IUBMB Life* **1998**, *45*, 923–930. [CrossRef]
21. Skibola, C.F.; Smith, M.T. Potential health impacts of excessive flavonoid intake. *Free. Radic. Biol. Med.* **2000**, *29*, 375–383. [CrossRef]
22. Rader, B.A. Alkaline phosphatase, an unconventional immune protein. *Front. Immunol.* **2017**, *8*, 897. [CrossRef]
23. Yan, J.; Luo, K.; Wang, B.; He, J.; Li, C.; Zhou, J.; Lai, W.; Zhou, X.; Ye, J.; Pan, M. Effects of dietary compound acidifier (Biomin) on the growth, anti-oxidation, immunity, and intestinal health of juvenile channel catfish (*Ictalurus punctatus*). *Aquac. Rep.* **2025**, *40*, 102577. [CrossRef]
24. Chipman, D.M.; Sharon, N. Mechanism of Lysozyme Action: Lysozyme is the first enzyme for which the relation between structure and function has become clear. *Science* **1969**, *165*, 454–465. [CrossRef]
25. Gaweł, S.; Wardas, M.; Niedworok, E.; Wardas, P. Malondialdehyde (MDA) as a lipid peroxidation marker. *Wiad. Lek.* **2004**, *57*, 453–455.
26. Jiang, M.; Li, Z.; Zhu, G. Immunological regulatory effect of flavonoid baicalin on innate immune toll-like receptors. *Pharmacol. Res.* **2020**, *158*, 104890. [CrossRef]
27. Peng-Fei, L.; Fu-Gen, H.; Bin-Bin, D.; Tian-Sheng, D.; Xiang-Lin, H.; Ming-Qin, Z. Purification and antioxidant activities of baicalin isolated from the root of huangqin (*Scutellaria baicalensis gcorsi*). *J. Food Sci. Technol.* **2013**, *50*, 615–619. [CrossRef]
28. Xi, Y.; Wu, M.; Li, H.; Dong, S.; Luo, E.; Gu, M.; Shen, X.; Jiang, Y.; Liu, Y.; Liu, H. Baicalin attenuates high fat diet-induced obesity and liver dysfunction: Dose-response and potential role of CaMKKβ/AMPK/ACC pathway. *Cell. Physiol. Biochem.* **2015**, *35*, 2349–2359. [CrossRef] [PubMed]
29. Elshopakey, G.E.; Mahboub, H.H.; Sheraiba, N.I.; Abduljabbar, M.H.; Mahmoud, Y.K.; Abomughaid, M.M.; Ismail, A.K. Ammonia toxicity in Nile tilapia: Potential role of dietary baicalin on biochemical profile, antioxidant status and inflammatory gene expression. *Aquac. Rep.* **2023**, *28*, 101434. [CrossRef]
30. Kim, K.T.; Jeon, G.H.; Cho, S.H.; Lim, S.G.; Kwon, M.G.; Yoo, J.H. Effects of dietary inclusion of various concentrations of *Scutellaria baicalensis Georgi* extract on growth, body composition, serum chemistry and challenge test of far eastern catfish (*Silurus asotus*). *Aquac. Res.* **2013**, *44*, 1502–1510. [CrossRef]

31. Mosconi-Bac, N. Hepatic disturbances induced by an artificial feed in the sea bass (*Dicentrarchus labrax*) during the first year of life. *Aquaculture* **1987**, *67*, 93–99. [CrossRef]
32. Wroblewski, F.; Ladue, J.S. Serum glutamic pyruvic transaminase in cardiac and hepatic disease. *Proc. Soc. Exp. Biol. Med.* **1956**, *91*, 569–571. [CrossRef]
33. Guo, H.-x.; Liu, D.-h.; Ma, Y.; Liu, J.-f.; Wang, Y.; Du, Z.-Y.; Wang, X.; Shen, J.-K.; Peng, H.-L. Long-term baicalin administration ameliorates metabolic disorders and hepatic steatosis in rats given a high-fat diet. *Acta Pharmacol. Sin.* **2009**, *30*, 1505–1512. [CrossRef]
34. Liu, J.W.; Li, X.Q.; You, Y.H.; Cheng, Z.H.; Liu, H.N.; Leng, X.J. Effects of dietary sodium butyrate supplementation on growth performance, serum biochemical and antioxidant indices, and intestinal health of juvenile channel catfish (*Ictalurus punctatus*). *Chin. J. Anim. Nutr.* **2024**, *36*, 5895–5909.
35. Wang, X.; Xie, L.; Long, J.; Liu, K.; Lu, J.; Liang, Y.; Cao, Y.; Dai, X.; Li, X. Therapeutic effect of baicalin on inflammatory bowel disease: A review. *J. Ethnopharmacol.* **2022**, *283*, 114749. [CrossRef] [PubMed]
36. Zhang, H.; Wang, X.; Qu, P.; Huang-Fu, Y.-X.; Liu, D.; Wu, Y.; Liu, Y.; Chen, P.; Mai, K.; Zhang, W. Herbs mixture improves growth performance, intestine and liver histology, and immunity of juvenile large yellow croaker (*Larimichthys crocea*). *Aquac. Rep.* **2024**, *36*, 102136. [CrossRef]
37. O'Hara, A.M.; Shanahan, F. The gut flora as a forgotten organ. *EMBO Rep.* **2006**, *7*, 688–693. [CrossRef]
38. Simon, G.L.; Gorbach, S.L. Intestinal flora in health and disease. *Gastroenterology* **1984**, *86*, 174–193. [CrossRef]
39. Chen, Y.X.; Wang, G.X.; Qiu, J.Q.; Liu, Y.; Shi, H.Q.; Dong, R.Q.; Peng, K. Curcumin alleviates the adverse effects of oxidized fish oil on the intestine of yellow catfish (*Pelteobagrus fulvidraco*). *Chin. J. Anim. Nutr.* **2024**, *36*, 5297–5307.
40. Egerton, S.; Culloty, S.; Whooley, J.; Stanton, C.; Ross, R.P. The gut microbiota of *Marine fish*. *Front. Microbiol.* **2018**, *9*, 873. [CrossRef] [PubMed]
41. Bennett, K.; Eley, A. Fusobacteria: New taxonomy and related diseases. *J. Med. Microbiol.* **1993**, *39*, 246–254. [CrossRef]
42. Magnúsdóttir, S.; Ravcheev, D.; de Crécy-Lagard, V.; Thiele, I. Systematic genome assessment of B-vitamin biosynthesis suggests co-operation among gut microbes. *Front. Genet.* **2015**, *6*, 148. [CrossRef]
43. Zhu, H.; Qiang, J.; He, J.; Tao, Y.; Bao, J.; Xu, P. Physiological parameters and gut microbiome associated with different dietary lipid levels in hybrid yellow catfish (*Tachysurus fulvidraco*♀× *Pseudobagrus vachellii*♂). *Comp. Biochem. Physiol. Part D Genom. Proteom.* **2021**, *37*, 100777. [CrossRef] [PubMed]
44. Ringø, E.; Gatesoupe, F.-J. Lactic acid bacteria in fish: A review. *Aquaculture* **1998**, *160*, 177–203. [CrossRef]
45. Gallet, A.; Halary, S.; Duval, C.; Huet, H.; Duperron, S.; Marie, B. Disruption of fish gut microbiota composition and holobiont's metabolome during a simulated Microcystis aeruginosa (*Cyanobacteria*) bloom. *Microbiome* **2023**, *11*, 108. [CrossRef]
46. Xia, Y.-T.; Cheng, E.H.-C.; Wang, H.-Y.; Zhang, L.H.-L.; Lin, S.-Y.; Dong, T.T.-X.; Duan, R.; Qin, Q.-W.; Wang, W.-X.; Tsim, K.W.-K. The extract from aerial part of *Scutellaria baicalensis* regulates gut microbiota in rabbit fish: Replacement of antibiotic fighting against pathogenic bacteria. *Aquaculture* **2023**, *565*, 739140. [CrossRef]
47. Du, Y.; Han, Y.; Zhang, R.; Zhang, Y.; Bao, S.; Cao, Y. Dietary baicalein improves growth performance, antioxidant activity, and intestinal flora of koi carp (*Cyprinus carpio*). *Aquac. Rep.* **2022**, *27*, 101421. [CrossRef]
48. Jiang, Y.; Xiong, X.; Danska, J.; Parkinson, J. Metatranscriptomic analysis of diverse microbial communities reveals core metabolic pathways and microbiome-specific functionality. *Microbiome* **2016**, *4*, 2. [CrossRef] [PubMed]
49. Wong, J.M.; De Souza, R.; Kendall, C.W.; Emam, A.; Jenkins, D.J. Colonic health: Fermentation and short chain fatty acids. *J. Clin. Gastroenterol.* **2006**, *40*, 235–243. [CrossRef]
50. Sun, Y.; O'Riordan, M.X. Regulation of bacterial pathogenesis by intestinal short-chain fatty acids. *Adv. Appl. Microbiol.* **2013**, *85*, 93–118.
51. Ju, M.; Liu, Y.; Li, M.; Cheng, M.; Zhang, Y.; Deng, G.; Kang, X.; Liu, H. Baicalin improves intestinal microecology and abnormal metabolism induced by high-fat diet. *Eur. J. Pharmacol.* **2019**, *857*, 172457. [CrossRef]
52. August, P.R.; Tang, L.; Yoon, Y.J.; Ning, S.; Müller, R.; Yu, T.-W.; Taylor, M.; Hoffmann, D.; Kim, C.-G.; Zhang, X. Biosynthesis of the ansamycin antibiotic rifamycin: Deductions from the molecular analysis of the rif biosynthetic gene cluster of *Amycolatopsis mediterranei* S699. *Chem. Biol.* **1998**, *5*, 69–79. [CrossRef] [PubMed]
53. Griffith, R.S. Introduction to vancomycin. *Rev. Infect. Dis.* **1981**, *3*, S200–S204. [CrossRef] [PubMed]
54. Yang, L.; Bajinka, O.; Jarju, P.O.; Tan, Y.; Taal, A.M.; Ozdemir, G. The varying effects of antibiotics on gut microbiota. *AMB Express* **2021**, *11*, 116. [CrossRef] [PubMed]

50. Sheng, H.; Li, H.; Huang, Y.; Zhang, B.; Liang, J.; Zhou, X.; Tian, Y.; Li, Q. The Sensing Selectivity of Gas Sensors Based on Different Sn-Doped Indium Oxide Films. *Chemosensors* **2025**, *13*, 169. [CrossRef]
51. Kita, J.; Engelbrecht, A.; Schubert, F.; Groß, A.; Rettig, F.; Moos, R. Some practical points to consider with respect to thermal conductivity and electrical resistivity of ceramic substrates for high-temperature gas sensors. *Sens. Actuators B Chem.* **2015**, *213*, 541–546. [CrossRef]
52. Zhang, R.; Wang, Y.; Zhang, Z.; Cao, J. Highly Sensitive Acetone Gas Sensor Based on g-C_3N_4 Decorated $MgFe_2O_4$ Porous Microspheres Composites. *Sensors* **2018**, *18*, 2211. [CrossRef]
53. Michel, G.; Martínez, A.H.; Jiménez, S. Gas sensing response of nanostructured trirutile-type $CoSb_2O_6$ synthesized by solution-polymerization method. *Sens. Actuators B* **2008**, *132*, 45–51. [CrossRef]
54. Wang, C.; Yin, L.; Zhang, L.; Xiang, D.; Gao, R. Metal Oxide Gas Sensors: Sensitivity and Influencing Factors. *Sensors* **2010**, *10*, 2088–2106. [CrossRef]
55. Rama Murthy, S.A.; Prabhu, E.; Clinsha, P.C.; Ravindranath Nair, A.; Gnanasekar, K.I.; Jayaraman, V. Role of oxygen species towards chemical sensing by semiconducting oxides. *Appl. Surf. Sci.* **2019**, *487*, 362–368.
56. Srivastava, J.K.; Pandey, P.; Mishra, V.N.; Dwivedi, R. Sensing mechanism of Pd-doped SnO_2 sensor for LPG detection. *Solid State Sci.* **2009**, *11*, 1602–1605. [CrossRef]
57. Yang, X.; Zhang, S.; Yu, Q.; Zhao, L.; Sun, P.; Wang, T.; Liu, F.; Yan, X.; Gao, Y.; Liang, X.; et al. One step synthesis of branched SnO_2/ZnO heterostructures and their enhanced gas-sensing properties. *Sens. Actuators B* **2019**, *281*, 415–423. [CrossRef]
58. Gildo-Ortiz, L.; Reyes-Gómez, J.; Flores-Álvarez, J.M.; Guillén-Bonilla, H.; Olvera, M.; Rodríguez Betancourtt, V.M.; Verde-Gómez, Y.; Guillén-Cervantes, A.; Santoyo-Salazar, J. Synthesis, characterization and sensitivity tests of perovskite-type $LaFeO_3$ nanoparticles in CO and propane atmospheres. *Ceram. Int.* **2016**, *42*, 18821–18827. [CrossRef]
59. Kapse, V.D. Preparation of Nanocrystalline Spinel-type oxide Materials for Gas sensing Applications. *Res. J. Chem. Sci.* **2015**, *5*, 7–12.
60. Wu, S.; Li, G.; Zhang, Y.; Zhang, W. Surface photoelectric and visible light driven photocatalytic properties of zinc antimonate-based photocatalysts. *Mater. Res. Bull.* **2013**, *48*, 1117–1121. [CrossRef]
61. Liu, W.; Lin, P.; Jin, H.; Xue, H.; Zhang, Y.; Li, Z. Nanocrystalline $ZnSb_2O_6$: Hydrothermal synthesis, electronic structure and photocatalytic activity. *J. Mol. Catal. A Chem.* **2011**, *349*, 80–85. [CrossRef]
62. Gnanaprakasam, A.; Sivakumar, V.M.; Thirumarimurugan, M. Influencing parameters in the photocatalytic degradation of organic effluent via nanometal oxide catalyst: A review. *Indian J. Mater.* **2015**, *2015*, 601827. [CrossRef]
63. Jabbar, Z.H.; Graimed, B.H. Recent developments in industrial organic degradation via semiconductor heterojunctions and the parameters affecting the photocatalytic process: A review study. *J. Water Process Eng.* **2022**, *47*, 102671. [CrossRef]
64. Charif, R.; Makhloufi, R.; Chaba Mouna, S.; Chadli, A.; Barkat, A.; Nouiri, M. Structural, electronic, optical, mechanical, and phononic bulk properties of tetragonal trirutile antimonate MSb_2O_6 (M = Fe, Co, Ni, or Zn): A first-principles prediction for optoelectronic applications. *Phys. Scr.* **2024**, *99*, 125909. [CrossRef]
65. Ham, K.; Hong, S.; Kang, S.; Cho, K.; Lee, J. Extensive active-site formation in trirutile $CoSb_2O_6$ by oxygen vacancy for oxygen evolution reaction in anion exchange membrane water splitting. *ACS Energy Lett.* **2021**, *6*, 364–370. [CrossRef]
66. Tran, H.D.; Nguyen, D.Q.; Do, P.T.; Tran, U.N.P. Kinetics of photocatalytic degradation of organic compounds: A mini-review and new approach. *RSC Adv.* **2023**, *13*, 16915. [CrossRef]
67. Shaikh, B.; Ali Bhatti, M.; Ahmed Shah, A.; Tahira, A.; Karim Shah, A.; Usto, A.; Aftab, U.; Bukhari, S.I.; Alshehri, S.; Shah Bukhari, S.D.U.; et al. Mn_3O_4@ZnO Hybrid Material: An Excellent Photocatalyst for the Degradation of Synthetic Dyes Including Methylene Blue, Methyl Orange and Malachite Green. *Nanomaterials* **2022**, *12*, 3754. [CrossRef]
68. Kaur, H.; Hippargi, G.; Pophali, G.R.; Bansiwal, A.K. Treatment methods for removal of pharmaceuticals and personal care products from domestic wastewater. In *Pharmaceuticals and Personal Care Products: Waste Management and Treatment Technology*; Elsevier: Amsterdam, The Netherlands, 2019; pp. 129–150.
69. Gouvea, C.A.K.; Wypych, F.; Moraes, S.G.; Duran, N.; Nagata, N.; Peralta-Zamora, P. Semiconductor-assisted photocatalytic degradation of reactive dyes in aqueous solution. *Chemosphere* **2000**, *40*, 433–440. [CrossRef] [PubMed]
70. Liu, Y.; Li, L.; Li, X.; Xu, Y.; Wu, D.; Sakthivel, T.; Guo, Z.; Zhao, X.; Dai, Z. Asymmetric tacticity navigates the localized metal spin state for sustainable alkaline/sea water oxidation. *Sci. Adv.* **2025**, *11*, eads0861. [CrossRef] [PubMed]

MDPI AG
Grosspeteranlage 5
4052 Basel
Switzerland
Tel.: +41 61 683 77 34

Animals Editorial Office
E-mail: animals@mdpi.com
www.mdpi.com/journal/animals

www.ingramcontent.com/pod-product-compliance
Lightning Source LLC
Chambersburg PA
CBHW062102220926
51850CB00010B/2865
* 9 7 8 3 7 2 5 8 5 4 9 5 0 *